企业级卓越人才培养（信息类专业集群）解决方案“十三五”规划教材

Android 项目式程序设计

天津滨海迅腾科技集团有限公司　主编

南开大学出版社
天　津

图书在版编目(CIP)数据

Android 项目式程序设计/天津滨海迅腾科技集团有限公司主编. —天津：南开大学出版社，2017.6(2023.8 重印)
ISBN 978-7-310-05325-4

Ⅰ. ①A… Ⅱ. ①天… Ⅲ. ①移动终端－应用程序－程序设计 Ⅳ. ①TN929.53

中国版本图书馆 CIP 数据核字(2017)第 015057 号

Android 项目式程序设计
Android XIANGMU SHI CHENGXU SHEJI

南开大学出版社出版发行
出版人:陈　敬
地址:天津市南开区卫津路 94 号　　邮政编码:300071
营销部电话:(022)23508339　营销部传真:(022)23508542
https://nkup.nankai.edu.cn

河北文曲印刷有限公司印刷　全国各地新华书店经销
2017 年 6 月第 1 版　　2023 年 8 月第 4 次印刷
260×185 毫米　16 开本　16.5 印张　393 千字
定价:50.00 元

如遇图书印装质量问题,请与本社营销部联系调换,电话:(022)23508339

企业级卓越人才培养（信息类专业集群）解决方案
“十三五”规划教材编写委员会

编　者：王新强　廉新宇　李树真　刘　涛　刘　屏　李肖霆
李沛荣　朱　彬

前 言

随着移动互联网时代的到来，智能手机在人们生活中扮演的角色也越来越重要。由于Android系统具有较强的开放性、数据同步性、兼容性等优点，因此从众多手机操作系统中脱颖而出，成为使用人数最多的手机系统。

本书以项目为基础，由易到难排列，最终以Android技术知识点为教学项目的形式展现给读者，使读者读完本书后，对项目中的Android应用程序开发具有全面的了解，并且具备一定的项目开发能力。

本书包括一个岗前准备和十一个项目，即Android开发环境以及工程的创建、Android应用界面设计、界面跳转和信息传递、应用资源的使用、数据持久化的操作、复杂数据的展示、Service服务、广播接收者的使用、传感器值的获取、网络编程等，循序渐进地讲述了Android项目开发所需要的知识和技能。通过本书的学习，读者可以熟练地使用Eclipse进行Android项目的开发，了解Android界面设计和信息传递等知识，掌握Android相关服务，数据存储，网络编程的技能，从而设计出稳定高效的应用程序。

本书每个项目都按照Android知识体系循序渐进地讲解。每个项目都设有学习目标、任务描述、任务技能点详解、任务实现、任务拓展和任务总结。此结构条理清晰、内容详细。任务实现与任务拓展可以将所学的理论知识充分的应用到实战中。本书的十二个项目较为基础，学习起来难度较小，读者易于全面掌握所学的知识技能点。

本书由王新强、廉新宇主编，李树真、刘涛、刘屏、李肖霆、李沛荣、朱彬参与编写，由王新强、廉新宇、朱彬负责全面内容的规划，刘屏负责统稿、编排。具体分工如下：项目一由李肖霆编写；项目二、三由刘涛编写；项目四、五、六由王新强编写；项目七、八由李树真、李沛荣编写；项目九、十、十一由廉新宇、朱彬编写；项目十由刘屏编写。

本书理论内容简明扼要、即学即用，实例操作讲解细致、步骤清晰，实现了理论与实践的结合。操作步骤后有相对应的效果图，便于读者直观、清晰地看到操作效果，牢记书中的操作步骤。本书能使读者在Android系统的学习过程中更加顺利，并为后期Android系统的进一步学习打下坚实的基础。

企业级卓越人才培养(信息类专业集群)解决方案简介

企业级卓越人才培养(信息类专业集群)解决方案(以下简称“解决方案”)是面向我国职业教育量身定制的应用型、技术技能型人才培养解决方案,以天津滨海迅腾科技集团技术研发为依托,联合国内职业教育领域相关行业、企业、职业院校共同研究与实践研发的科研成果。本解决方案坚持“创新产教融合协同育人,推进校企合作模式改革”的宗旨,消化吸收德国“双元制”应用型人才培养模式,深入践行“基于工作过程”的技术技能型人才培养,设立工程实践创新培养的企业化培养解决方案。在服务国家战略、京津冀教育协同发展、中国制造2025(工业信息化)等领域培养不同层次及领域的信息化人才。为推进我国教育现代化发挥应有的作用。

该解决方案由“初、中、高级工程师”三个阶段构成,集技能型人才培养方案、专业教程、课程标准、数字资源包(标准课程包、企业项目包)、考评体系、认证体系、教学管理体系、就业管理体系等于一体。采用校企融合、产学融合、师资融合的模式在高校内共建互联网学院、软件学院、工程师培养基地的方式,开展“卓越工程师培养计划”,开设系列“卓越工程师班”,“将企业人才需求标准、企业工作流程、企业研发项目、企业考评体系、企业一线工程师、准职业人才培养体系、企业管理体系引进课堂”,充分发挥校企双方特长,推动校企、校际合作,促进区域优质资源共建共享,实现卓越人才培养目标,达到企业人才培养及招录的标准。本解决方案已在全国近二十所高校开始实施,目前已形成企业、高校、学生三方共赢格局。未来五年将努力实现在年培养能力达到万人的目标。

天津滨海迅腾科技集团是以IT产业为主导的高科技企业集团,总部设立在北方经济中心——天津,子公司和分支机构遍布全国近20个省市,集团旗下的迅腾国际、迅腾科技、迅腾网络、迅腾生物、迅腾日化分属于IT教育、软件研发、互联网服务、生物科技、快速消费品五大产业模块,形成了以科技为原动力的现代科技服务产业链。集团先后荣获“全国双爱双评先进单位”“天津市五一劳动奖状”“天津市政府授予AAA级和谐企业”“天津市文明单位”“高新技术企业”“骨干科技企业”等近百项殊荣。集团多年中自主研发天津市科技成果2项,具备自主知识产权的开发项目数十余项。现为国家工业和信息化部人才交流中心“全国信息化工程师”项目联合认证单位。

目 录

项目一 初识 Android 开发 …… 1

技能点 1 Android 概述 …… 2
技能点 2 开发环境搭建 …… 6
技能点 3 Eclipse 项目结构 …… 18
技能点 4 Android 常用开发工具及用法 …… 19
技能点 5 Eclipse 快捷键 …… 23

项目二 Android 应用界面设计 …… 33

技能点 1 控件属性介绍 …… 35
技能点 2 基本布局 …… 40
技能点 3 Dialog 介绍 …… 43

项目三 界面跳转和信息传递 …… 57

技能点 1 Activity 介绍 …… 59
技能点 2 Intent 介绍 …… 61

项目四 规范应用资源 …… 75

技能点 1 应用资源 …… 77
技能点 2 数组资源 …… 78
技能点 3 颜色资源文件 …… 79
技能点 4 尺寸资源 …… 81
技能点 5 动画 …… 82
技能点 6 样式与主题 …… 83
技能点 7 国际化 …… 84
技能点 8 布局资源 …… 85

项目五 数据持久化操作 …… 97

技能点 1 SharedPreferences 概述 …… 99
技能点 2 读写 SD 卡 …… 101
技能点 3 SQLite 数据库简介及操作 …… 103

项目六 复杂数据展示 …… 122

技能点 1 Adapter 接口 …… 124
技能点 2 Spinner 功能与用法 …… 125

技能点 3 ListView 概述 …… 126
技能点 4 GridView 功能与用法 …… 127

项目七 图形图像 …… 142

技能点 1 Bitmap 和 BitmapFactory …… 144
技能点 2 逐帧动画 …… 145
技能点 3 补间动画 …… 148
技能点 4 属性动画 …… 151

项目八 Service 服务 …… 163

技能点 1 Service 概述 …… 164
技能点 2 服务通信 …… 169

项目九 广播接收者 …… 182

技能点 1 广播接收者 …… 183
技能点 2 广播的发送与接收 …… 186

项目十 内容提供者 …… 198

技能点 1 ContentProvider 数据共享 …… 199
技能点 2 ContentProvider 实例模型 …… 201
技能点 3 ContentProvider 管理操作 …… 203

项目十一 传感器 …… 215

技能点 1 传感器简介 …… 216
技能点 2 Sensor …… 220
技能点 3 Vibrator …… 221

项目十二 网络编程 …… 236

技能点 1 线程 …… 237
技能点 2 Socket …… 239
技能点 3 HTTP …… 240
技能点 4 Message 与 Handler …… 242
技能点 5 JSON …… 245

项目一　初识 Android 开发

通过实现 HelloWorld 项目，学习 Android 开发环境的搭建及项目开发步骤，了解 Android 发展历程及 Android 相关软件的安装及使用，在项目实现过程中：

- 掌握 Android 开发环境的搭建；
- 掌握 Android 项目开发步骤；
- 掌握 Android 模拟器的使用；
- 掌握 DDMS 的使用。

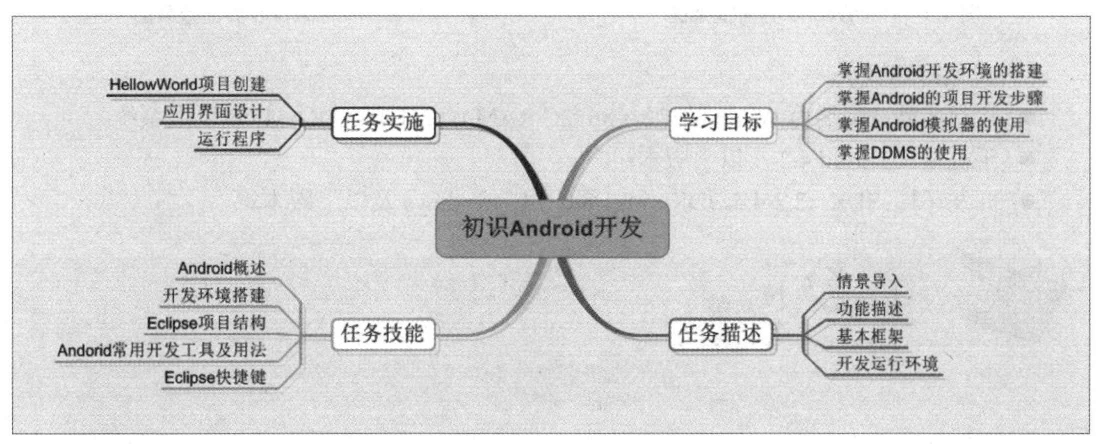

【情境导入】

Android 因拥有开放性、数据同步性、兼容性等优点，从手机操作系统中脱颖而出，成为使用人数最多的手机操作系统，其手机应用软件得以迅速发展。本次任务主要实现 HelloWorld

项目的创建、调试和运行。

【功能描述】

- 创建第一个 Android 项目 HelloWorld；
- 在虚拟机上运行项目；
- 使用 DDMS 调试。

【基本框架】

基本框架如图 1.1 所示。将框架图转换成的效果如图 1.2 所示。

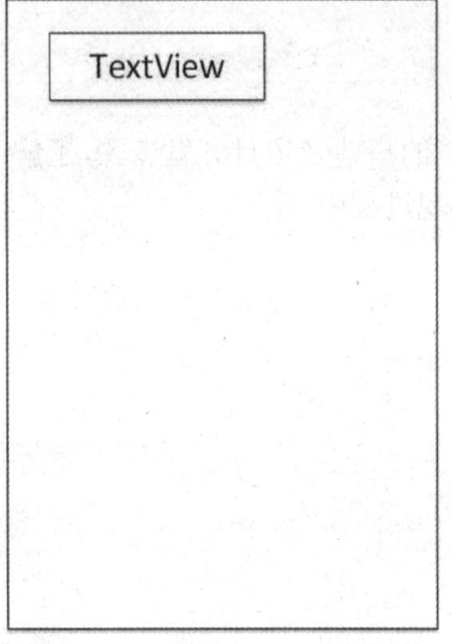

图 1.1　HelloWorld 界面框架图

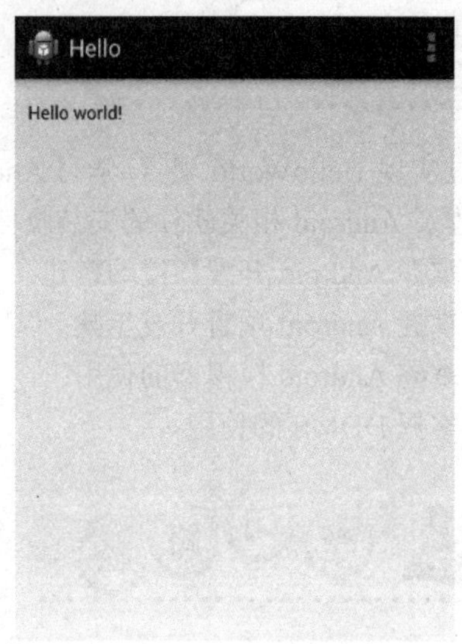

图 1.2　HelloWorld 界面效果图

【开发运行环境】

- 运行环境：PC 端为 Windows 7（×86，2G RAM）/（×64，4G RAM）及以上系统；
- 手机端：Android 4.2 及以上版本；
- 开发环境：SDK　2.2-4.4，JDK　jdk1.8.0_74 –windows 及以上版本。

技能点 1　Android 概述

1　Android 简介

Android 的本意是"机器人"，这个词汇最早出现于法国作家利尔亚当（Auguste Villiers de l'Isle-Adam）在 1886 年发表的科幻小说《未来夏娃》（L'ève future）中，小说中将外表像人的机器起名为 Android。

Android 是一个移动设备软件堆，包括操作系统、中间件、用户界面和关键应用软件。换言之，Android 是基于 Java 并运行在 Linux 内核上的轻量级操作系统，其功能覆盖面广泛，包括一系列 Google 公司在其中内置的应用软件，如打电话、发短信等基本应用功能。一个简单的 Android 运行界面如图 1.3 所示。

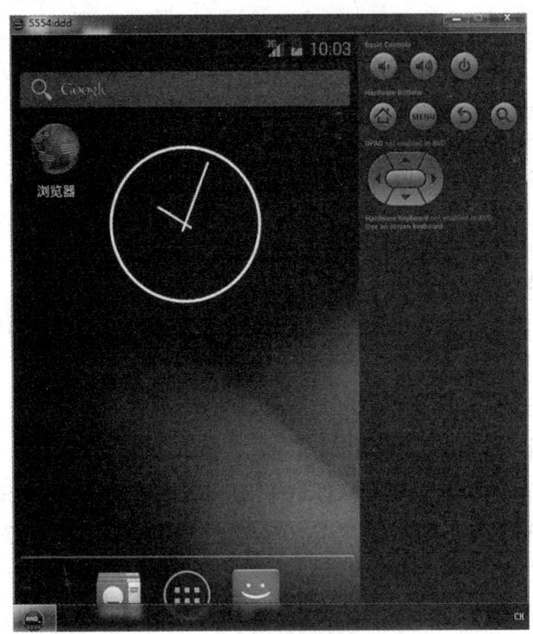

图 1.3　Android 运行界面

2　Android 版本

自 Android 首次发布至今，Android 已经出现很多的版本，如表 1.1 所示。

表 1.1　Android 版本列表

Android 版本	发布日期	代号
1.1	2009 年 02 月 09 日	Bender（发条机器人）
1.5	2009 年 04 月 30 日	Cupcake（纸杯蛋糕）
1.6	2009 年 09 月 15 日	Donut（炸面圈）
2.0/2.1	2009 年 10 月 26 日	Eclair(长松饼)
2.2	2010 年 05 月 20 日	Froyo（冻酸奶）
2.3	2010 年 12 月 06 日	Gingerbread（姜饼）
3.0	2011 年 02 月 03 日	Honeycmb（蜂巢）
4.1	2012 年 06 月 28 日	Jelly Bean（果冻豆）
4.2	2012 年 10 月 30 日	Jelly Bean（果冻豆）

续表

Android 版本	发布日期	代号
4.3	2013 年 07 月 25 日	Jelly Bean（果冻豆）
4.4	2013 年 11 月 01 日	KitKat（巧克力棒）
5.0/5.1	2014 年 10 月 16 日	Lollipop（棒棒糖）
6.0	2015 年 05 月 28 日	Marshmallow（棉花糖）
7.0	2016 年 05 月 18 日	Nougat（牛轧糖）

3　Android 功能

Android 的功能强大，具体包括以下几个功能：

- 存储：使用 SQLite（轻量级的关系数据库）进行数据存储
- 连接性：支持 GSM/EDGE、IDEN、CDMA、EV-DO、UMTS、Bluetooth（包括 A2DP 和 AVRCP）、Wi-Fi、LTE 和 WiMAX；
- 消息传递：支持 SMS 和 MMS；
- Web 浏览器：基于开源的 WebKit，并集成 Chrome 的 V8 JavaScript 引擎；
- 媒体支持：支持以下媒体：H.263、H.264（在 3GP 或 MP4 容器中）、MPEG-4 SP、AMR、AMR-WB、AAC、HE-AAC（在 MP4 或 3GP 容器中）、MP3、MIDI、WAV、JPEG、PNG、GIF 和 BMP；
- 硬件支持：加速传感器、摄像头、数字式罗盘、接近传感器和全球定位系统；
- 多点触摸：支持多点触摸屏幕；
- 多任务：支持多任务应用；
- Flash 支持：Android 3.0 及以上版本支持 Flash 10.1。

4　Android 架构

Android 操作系统的各个层面如图 1.4 所示。通过对 Android 架构的学习，使读者更全面地了解 Android 系统。

从架构图看，Android 分为四个层，从高层到低层分别是应用程序（Application）层、应用程序框架（Application Framework）层、系统运行库（Librares）层和 Linux 内核（Linux Kernel）层。

应用程序层：主要是 Android 自带的一些应用程序，例如：电话、联系人、浏览器等，还包括从 Android Market 应用程序商店下载和安装的应用程序。

应用程序框架层：主要是对程序员开放的 Android 操作系统的各种功能，以便在应用程序中使用各项功能。

系统运行库层：主要包含一些 C/C++ 库，这些库能被 Android 系统中不同的组件使用。

Linux 内核层：为 Android 的内核，包括 Android 设备的各种硬件组建的底层设备驱动程序。

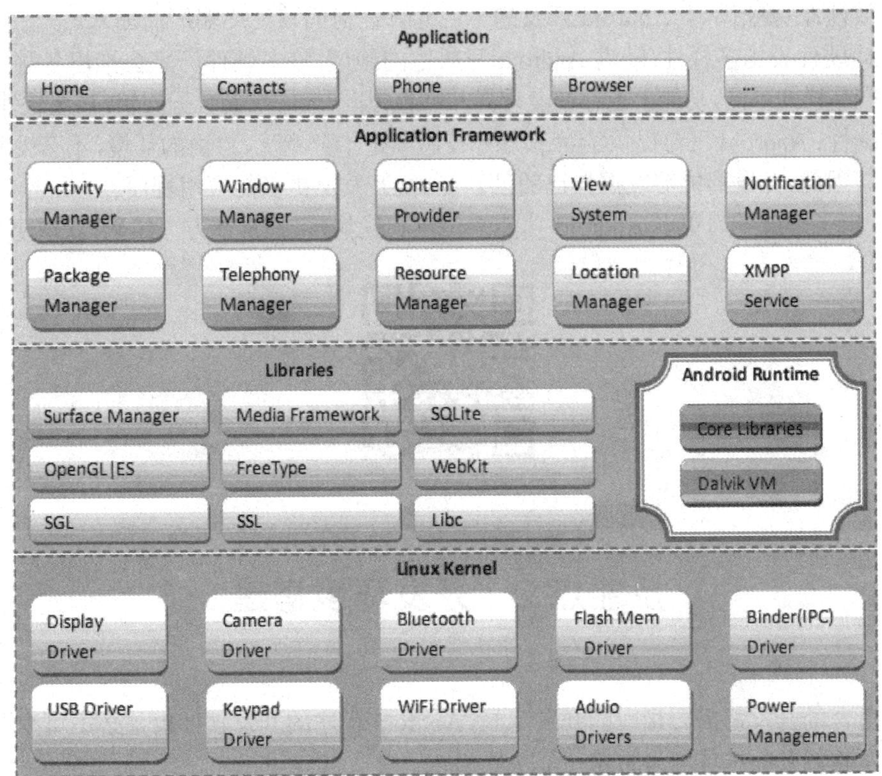

图 1.4　Android 操作系统(OS)的各个层面

5　Android 特性

Android 具有如下特性：

- 能够灵活地运用程序框架，支持组件的重用和替换；
- 娱乐功能丰富，包括常见的音频、视频和静态映像文件格式（如 MPEG4、MP3、AAC、AMR、JPG、PNG 和 GIF）；
- 优化的图形库，包括定制 2D 图形库和 3D 图形库，其中 3D 图形库基于 OpenGL ES 1.0；
- 拥有专门的为移动设备优化的虚拟机——Dalvik；
- 内部集成浏览器，这个浏览器是基于开源的 WebKit 引擎；
- 结构化的数据存储使用了 SQLite 数据库；
- 支持 USB、蓝牙、Wi-Fi 等多种数据传输（依赖于硬件）；
- 支持摄像头、GPS、指南针和加速度器（依赖于硬件）；
- 丰富的开发环境，包括设备模拟器、调试工具、内存及性能分析图表和 Eclipse 集成开发环境插件等；
- 支持 GSM、WCDM、LTE 等多种移动电话技术。

6　Android 优势

Android 和其他编程语言相比，具有以下优势：

- 开放性：主要指基于 Android 开发的平台允许任何的移动终端厂商加入。
- 支持硬件设施多样性：随着 Android 开放性的施展，许多硬件厂家会推出各种不同的产品。尽管产品样式不同，功能上也存在着差异和特色，却不会影响到数据同步，甚至软件的兼容。
- 便捷性：Android 平台提供给第三方开发商一个十分宽泛、自由的环境，不会受到各种规定的束缚。因此，开发商能够发挥自己的创新能力，开发出更多的应用程序。

拓展：想了解或学习更多 Android 功能知识点和优势，可扫描下方二维码，获取更多信息。

技能点 2　开发环境搭建

每种语言的开发都需要相应的开发工具，Android 程序的开发软件也是必不可少的。本书采用的开发软件及版本环境为 JDK 1.8 以及 SDK。

1　JDK(Java Development Kit)

JDK 是 Java 语言的软件开发工具包，主要用于移动设备、嵌入式设备上的 Java 应用程序开发，是搭建 Java 开发运行环境最基本要素。JDK 中包含一些开发所需要的工具的集合。

2　Eclipse

Eclipse 是跨平台自由集成开发环境（IDE），是一个框架平台。Eclipse 本身是一个框架和一组服务，可通过插件组件构建开发环境。Eclipse 附带了标准的插件集，其中包括 Java 开发工具（Java Development Tools，JDT）。Eclipse 开发界面如图 1.5 所示。

3　Android SDK

SDK 是 Software Development Kit 的缩写，是软件工程师为所使用的特定的软件包、软件框架、硬件平台、操作系统等建立应用软件开发工具的集合，而 Android SDK 指的是 Android 专属的软件开发工具包。Android SDK 不用安装，下载后将 SDK 压缩包解压即可。

4　ADT

ADT（Android Development Tools）是 Android 为 Eclipse 所定制的一个插件，这个插件的主要作用是为用户开发 Android 应用程序，提供一个强大的综合环境。它拓展了 Eclipse 的功能，可以让用户快速地建立 Android 项目，创建应用程序界面，在基于 Android 框架 API 的基础上添加组件，以及用 SDK 工具集调试应用程序，甚至导出签名（或未签名）的 APK 以便运

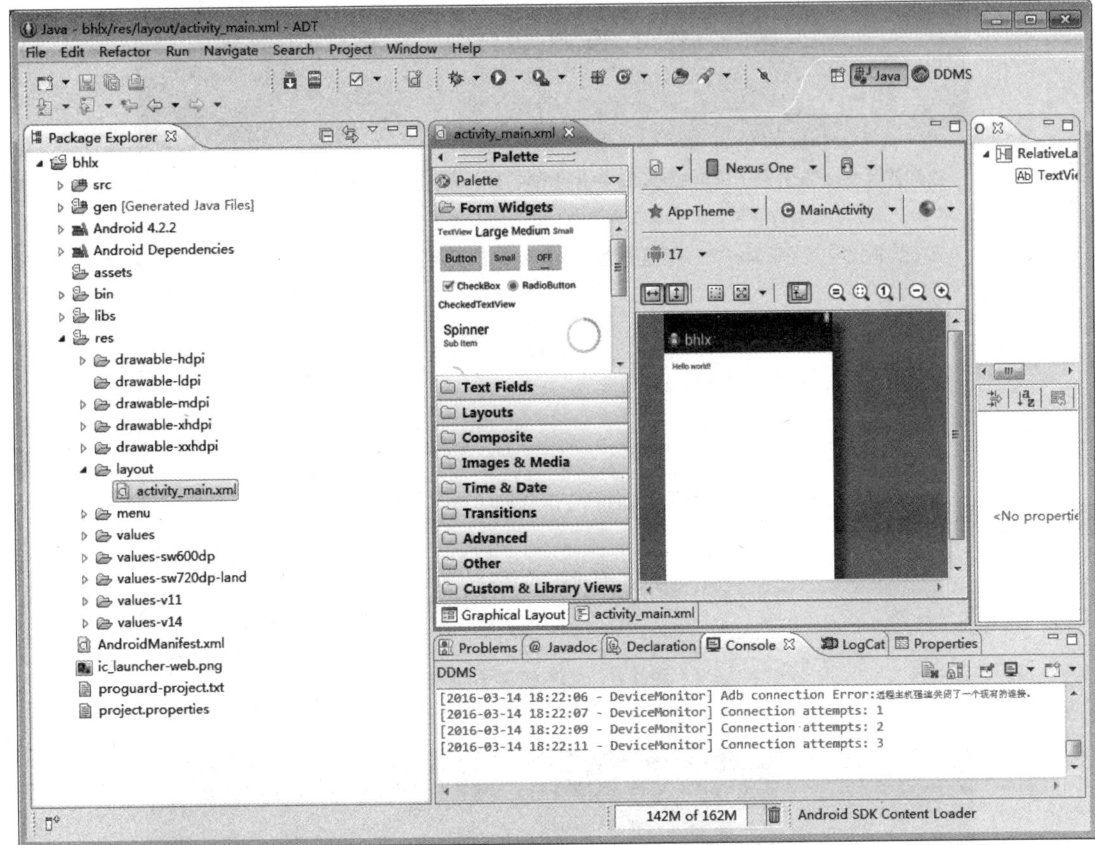

图 1.5　Eclipse 开发界面

行应用程序。

5　安装步骤

第一步：安装 JDK（Java Development Kit）。

首先运行该程序，然后根据提示来选择安装路径，将 JDK 安装到指定的文件夹即可，一般情况下是使用默认路径，具体步骤如下：

（1）首先打开安装界面，如图 1.6 所示。

（2）根据界面提示选择“下一步”，在这里我们选择“开发工具”一项，同时这里也可以选择其他安装路径。选择下一步，如图 1.7 所示。

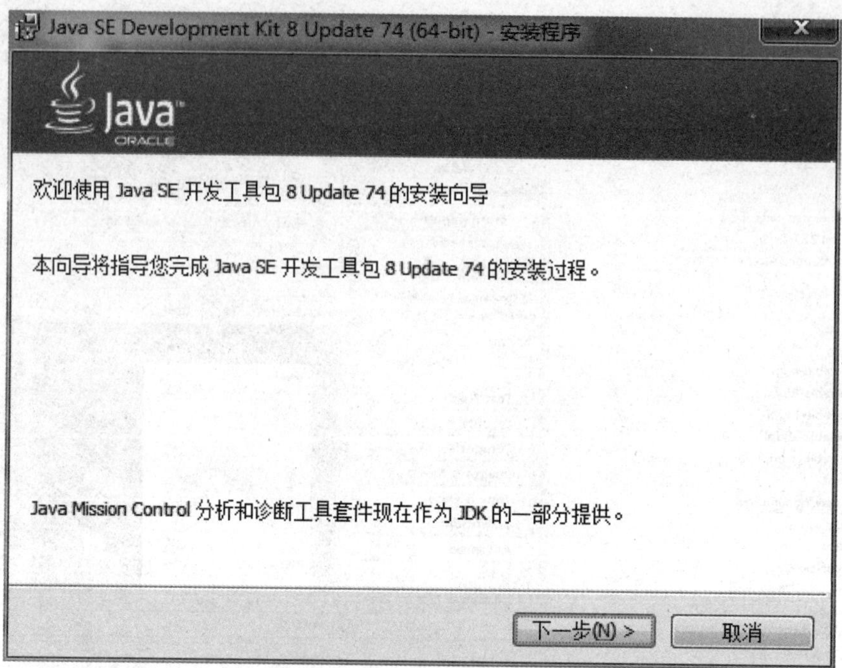

图 1.6 JDK 安装界面

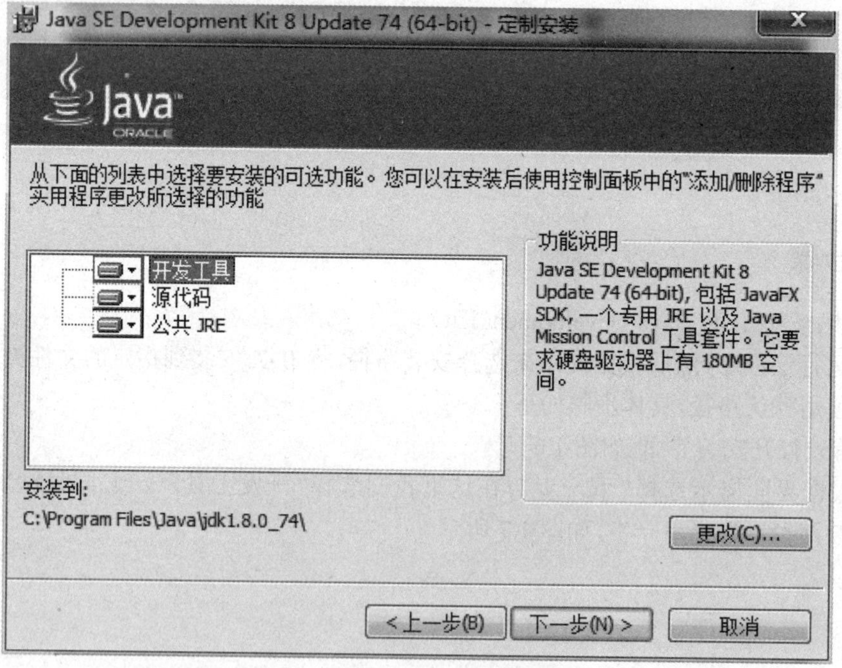

图 1.7 JDK 功能选项界面

(3)选择“下一步”,JDK 安装成功,如图 1.8 和图 1.9 所示。

图 1.8　安装界面

图 1.9　安装成功界面

（4）配置 JDK。选择“计算机”→“属性”→“高级系统设置”→“高级”→“环境变量”。如图 1.10 所示。

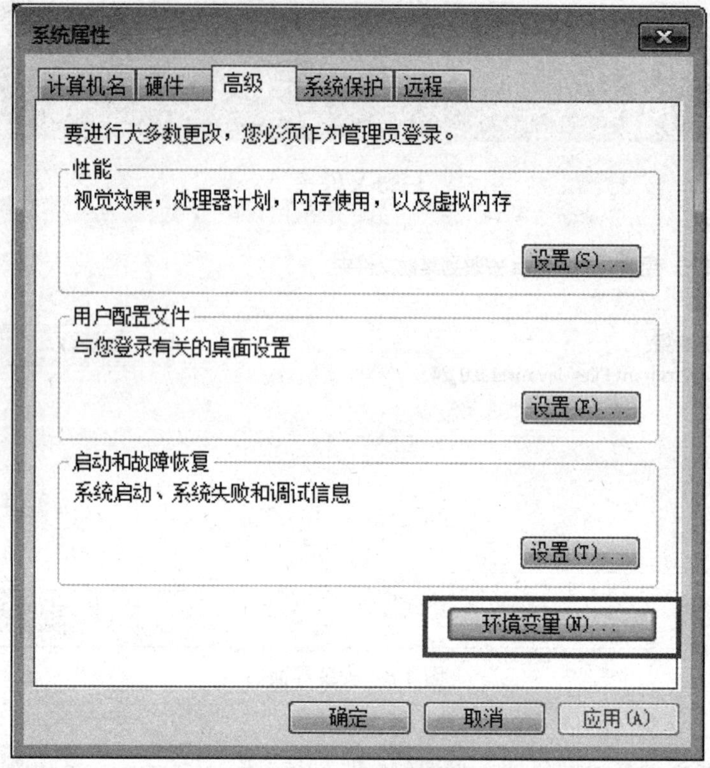

图 1.10 系统属性界面

(5)选择“系统变量”→新建“ JAVA_HOME”变量，变量值填写安装的 JDK 所在的位置路径。如图 1.11 所示。

图 1.11 JDK 位置路径

(6)“系统变量”→寻找“Path”变量“编辑”→在变量值最后输入:“%JAVA_HOME%\bin;%JAVA_HOME%\jre\bin;”。如图 1.12 所示。

图 1.12 系统变量设置

注意：原来 Path 的变量值末尾没有“;”号，先输入“;”号再输入上面的内容。

（7）“系统变量”→新建“CLASSPATH 变量”，变量值填写：

.;%JAVA_HOME%\lib;%JAVA_HOME%\lib\tools.jar

如图 1.13 所示。

注意：变量值开头有点。

图 1.13　系统变量设置

（8）检测其是否已经配置成功，检测的步骤是：“开始”→“运行”命令，在“运行”对话框的文本框中输入“cmd”，在打开的 CMD 窗口中输入 java -version。如果显示如图 1.14 所示的提示信息，则说明 JDK 安装成功。

注意：java 和 -version 之间是有空格的。

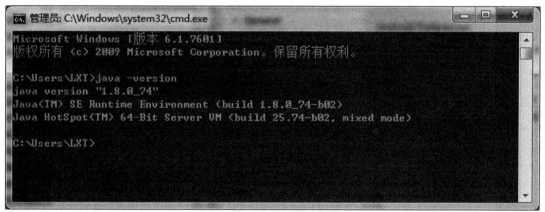

```
管理员: C:\Windows\system32\cmd.exe
Microsoft Windows [版本 6.1.7601]
版权所有 (c) 2009 Microsoft Corporation。保留所有权利。

C:\Users\LXT>java -version
java version "1.8.0_74"
Java(TM) SE Runtime Environment (build 1.8.0_74-b02)
Java HotSpot(TM) 64-Bit Server VM (build 25.74-b02, mixed mode)

C:\Users\LXT>
```

图 1.14　命令窗口

第二步：安装 Eclipse。

下载最新版本的 Eclipse 集成开发环境，具体步骤如下：

（1）将下载的 Eclipse 安装文件解压到硬盘上的某个目录，如图 1.15 所示。

（2）Eclipse 集成开发环境是无需安装的，在解压并打开 Eclipse 后，找到用户安装的 JDK 路径进入解压后的目录。双击可执行文件“eclipse.exe”，运行 Eclipse，出现如图 1.16 所示界面，选择文件工作空间路径，点击确定。

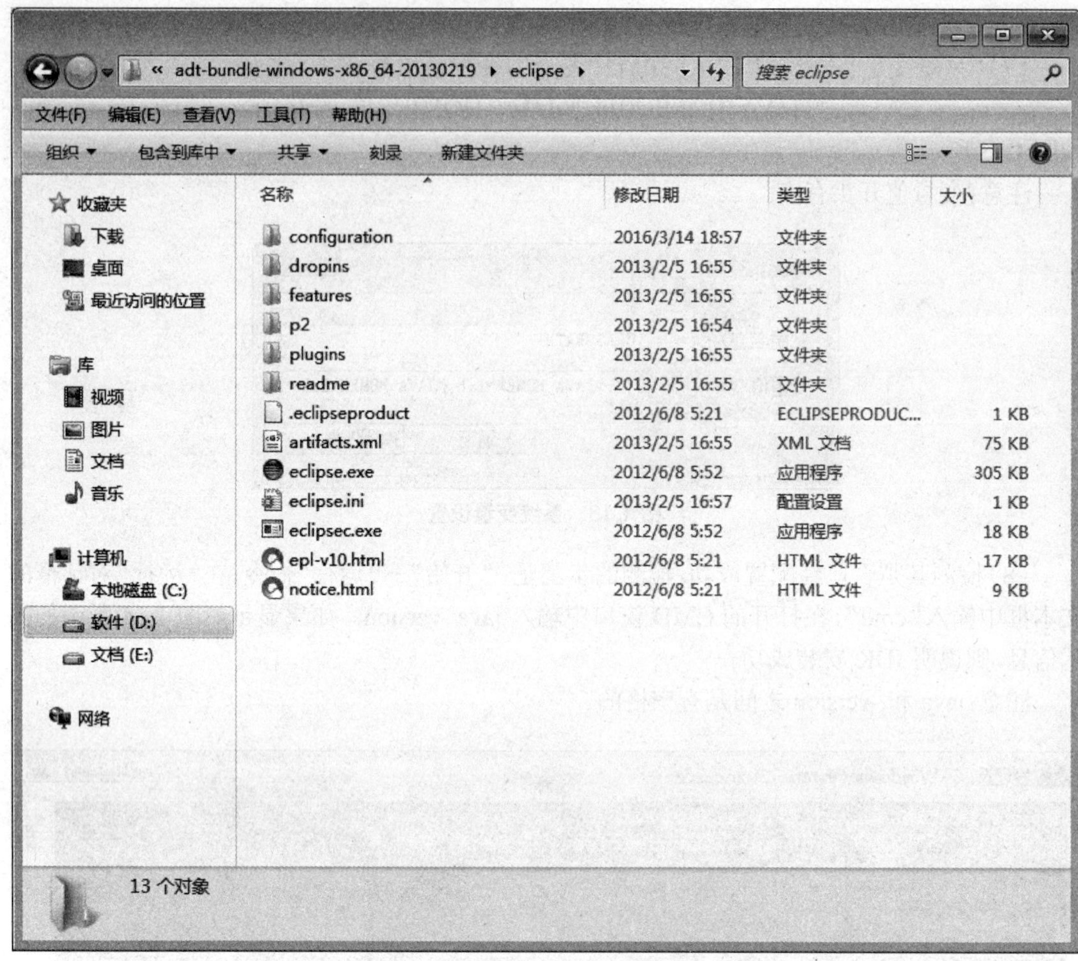

图 1.15 解压后的 Eclipse 目录

Workspace Launcher
Select a workspace
ADT stores your projects in a folder called a workspace.
Choose a workspace folder to use for this session.
Workspace: E:\anzhuo Browse...
Use this as the default and do not ask again
OK Cancel

图 1.16 选择 workspace 的界面

（3）点击“OK”，出现 Eclipse 操作界面如图。1.17 所示。

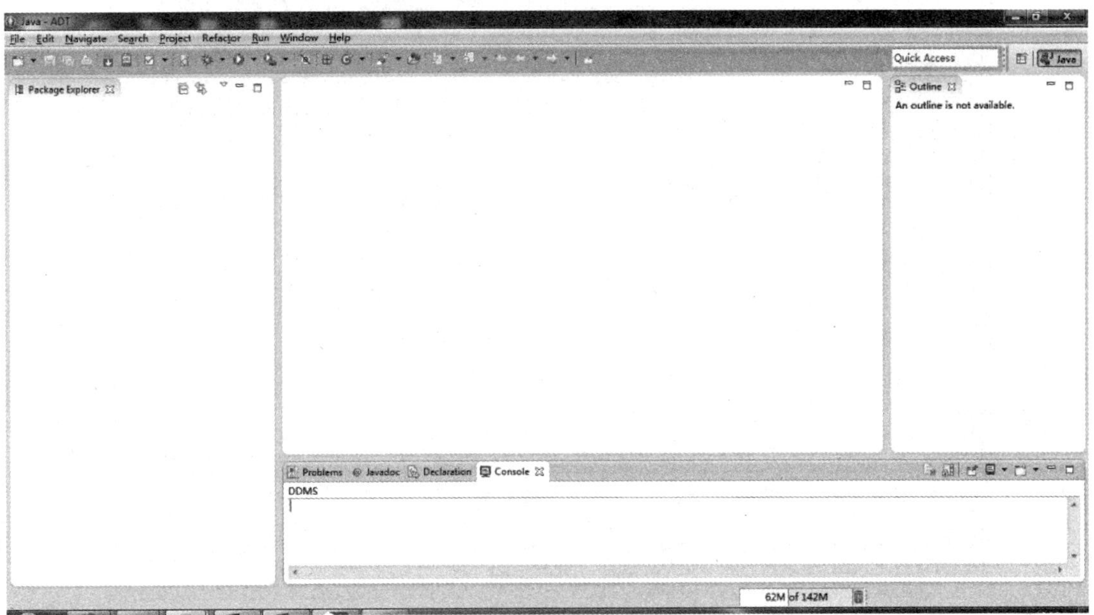

图 1.17　Eclipse 的操作界面

第三步：安装 Android SDK。

Android SDK 的安装，具体步骤如下：

（1）首先将现有的 Android SDK 开发包解压到某个盘的某个目录下。解压文件后，会得到以下几个重要的文件，但是在这里只选择“SDK Manager.exe”（负责下载和更新 SDK 包）。如图 1.18 所示。

名称	修改日期	类型	大小
add-ons	2012/6/25 4:53	文件夹	
platforms	2012/6/25 4:53	文件夹	
tools	2012/6/25 4:53	文件夹	
AVD Manager	2012/6/25 4:53	应用程序	350 KB
SDK Manager	2012/6/25 4:53	应用程序	350 KB
SDK Readme	2012/6/25 4:53	文本文档	2 KB

图 1.18　解压目录

（2）自动检测是否有更新的 SDK 数据包可供下载，然后选择所需安装的 Android 版本，然后点击“Install packages”安装。如图 1.19 所示。

（3）Android SDK 管理器开始下载并安装所选的包。如图 1.20 所示。

（4）安装完成后，在 Android SDK 管理器界面上，你所选的包在 Status 中会显示“Installed”，表示已经安装完成。如图 1.21 所示。

（5）将 SDK tools 目录的完整路径设置到系统变量中。新建变量名为“SDK_HOME”，在“变量值”文本框输入的 Android SDK 的解压目录的路径。如图 1.22 所示。

（6）找到“Path”的变量，点击编辑，在“变量值”文本框最前面加上“%SDK_HOME%\tools;”。如图 1.23 所示。

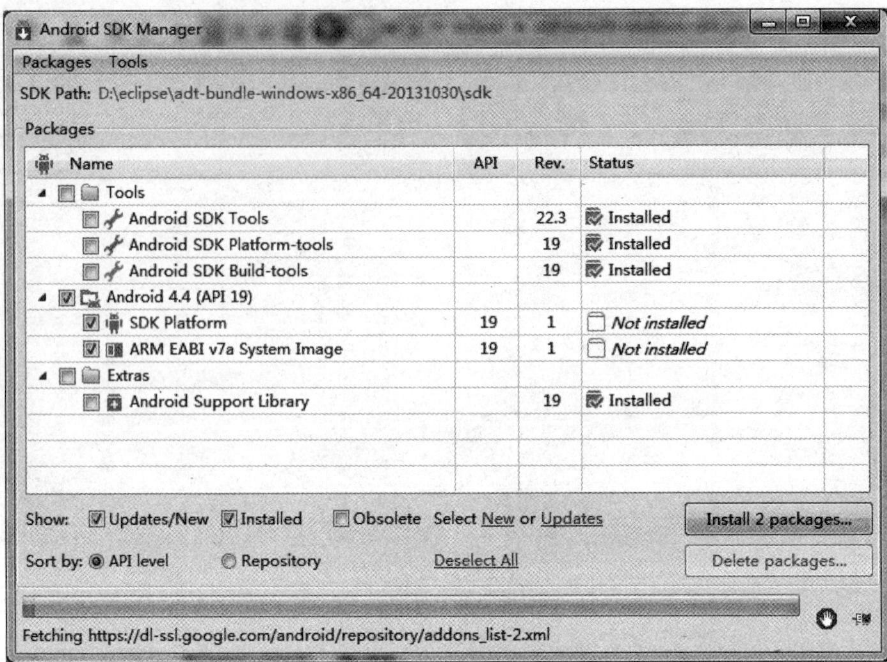

图 1.19 安装界面

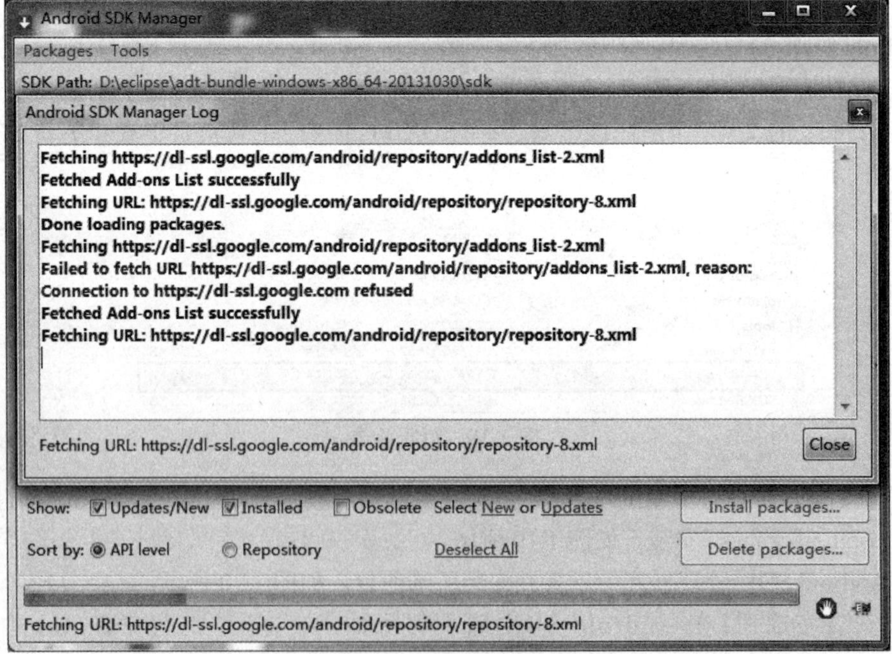

图 1.20 安装进程

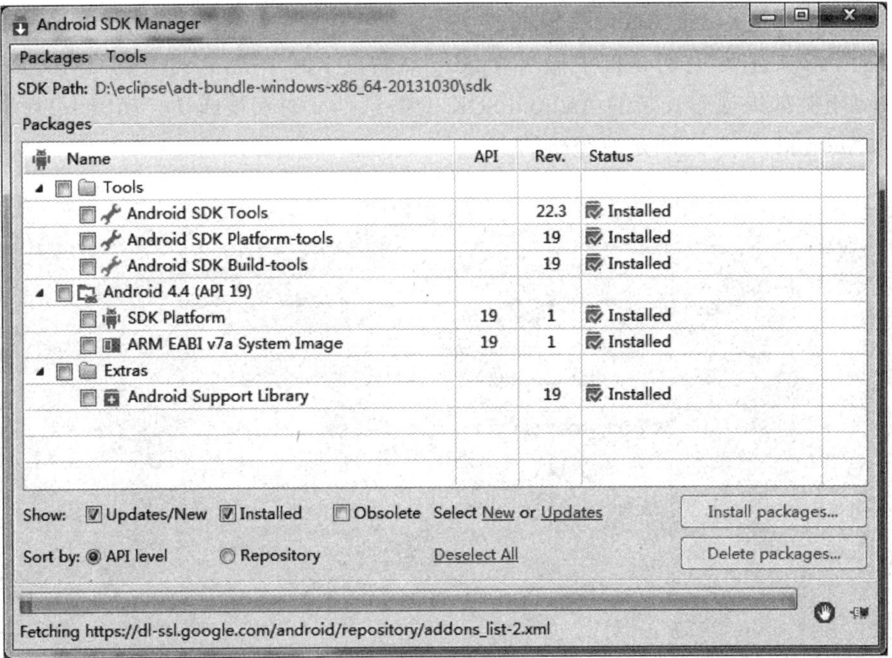

图 1.21　安装界面

图 1.22　设置系统变量

图 1.23　设置系统变量

（7）设置完成后，检查 Android SDK 是否已经安装成功，能够正常运行。依次单击“开始”→“运行”，然后在运行对话框中输入“cmd”，然后按下回车键，再打开的 CMD 命令窗口输入“Android -h”，如果显示安装的 Android SDK 的信息则证明安装成功。如图 1.24 所示。

注意：Android 和 -h 之间是有空格的。

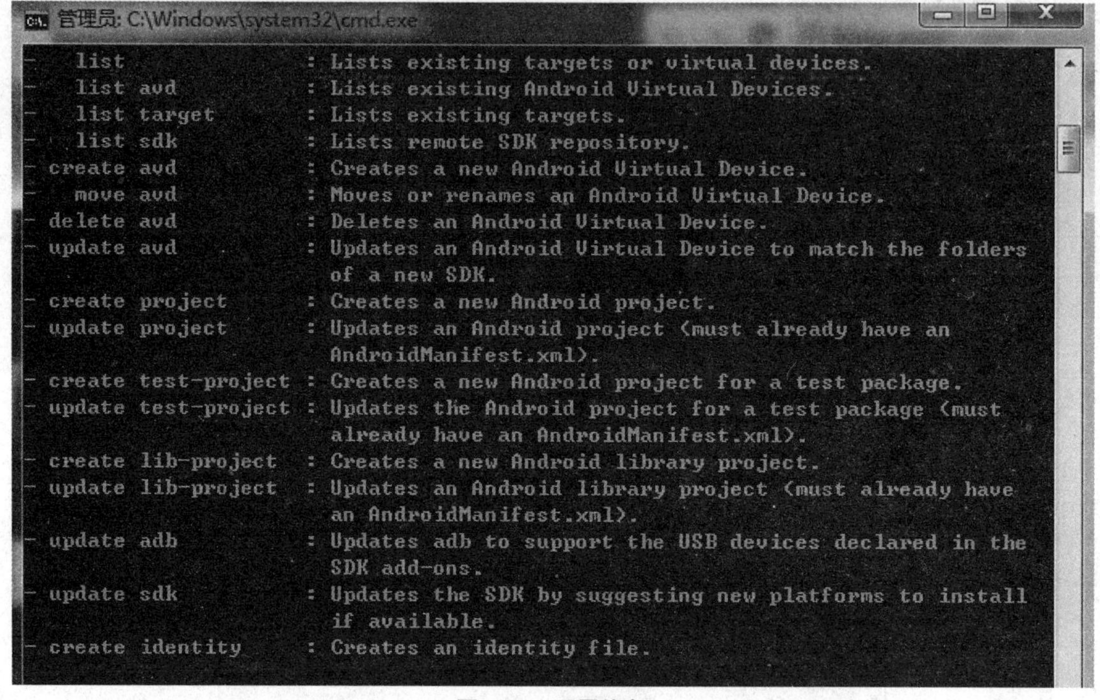

图 1.24 配置信息

第四步：将 ADT 和 Eclipse 绑定。

Android Development Tools（ADT）是 Android 为 Eclipse 定制的一个插件，该插件为用户提供一个强大的 Android 应用程序综合开发环境。ADT 是对 Eclipse IDE 的扩展，用来支持 Android 应用程序的创建和调试。安装步骤如下所示：

（1）在线安装

● 首先运行 Eclipse，启动 Eclipse IDE，选择菜单选项“Help”→“Install New Software”，出现“Install”界面，点击 Add 按钮，如图 1.25 所示。

● 在“name”文本框中输入名字，如：ADT Plugin。注意在“Location”文本框中不能再随意写地址，一定要输入插件的网络地址“https://dl-ssl.goole.com/Android/eclipse”，单击“OK”按钮。如图 1.26 所示。

第五步：设定 Android SDK 主目录。

插件安装后，在 Eclipse 中设置 Android SDK 的主目录，具体步骤如下：

（1）启动 Eclipse，在菜单中依次单击“Window”→“Preferences”，如图 1.27 所示。

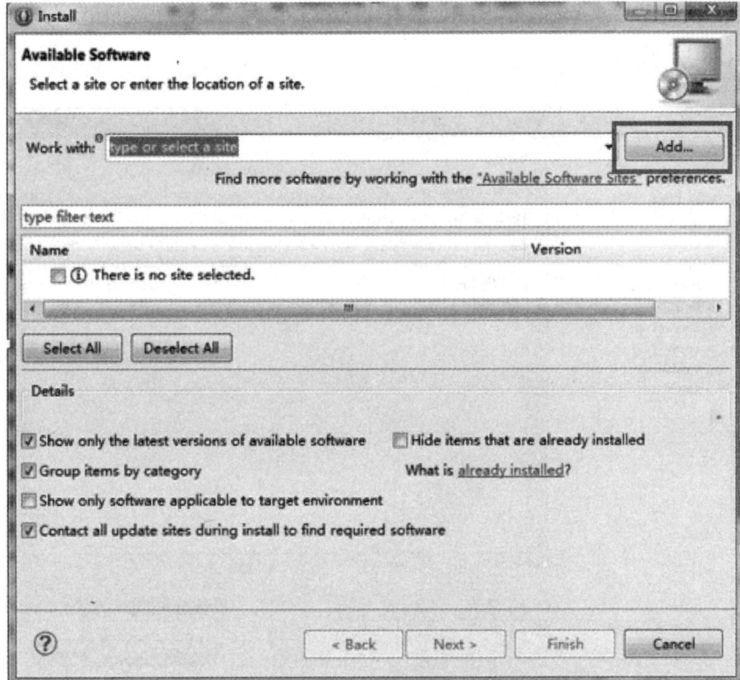

图 1.25　Install 对话框

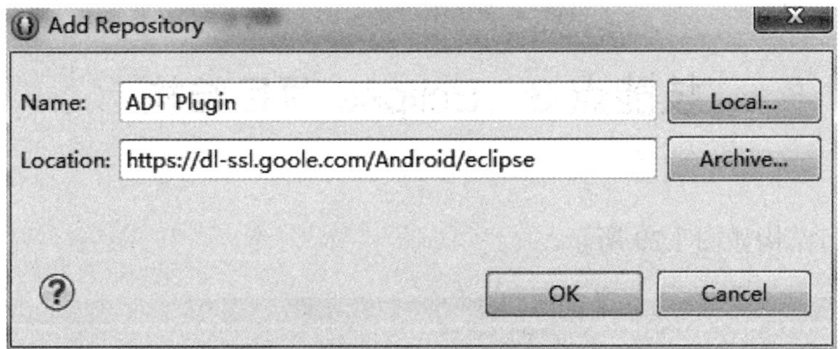

图 1.26　Add 对话框

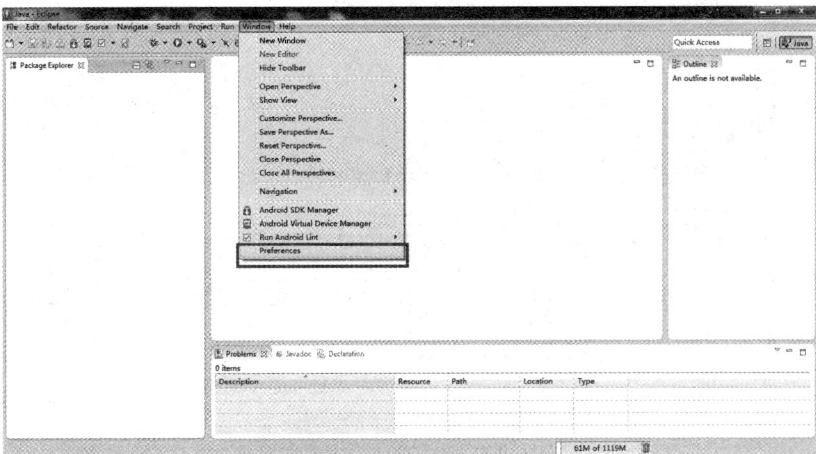

图 1.27　Eclipse 界面

（2）选择“Android”，将 Android SDK 所在目录设定为“SDK Location”，单击“OK”按钮完成设置。如图 1.28 所示。

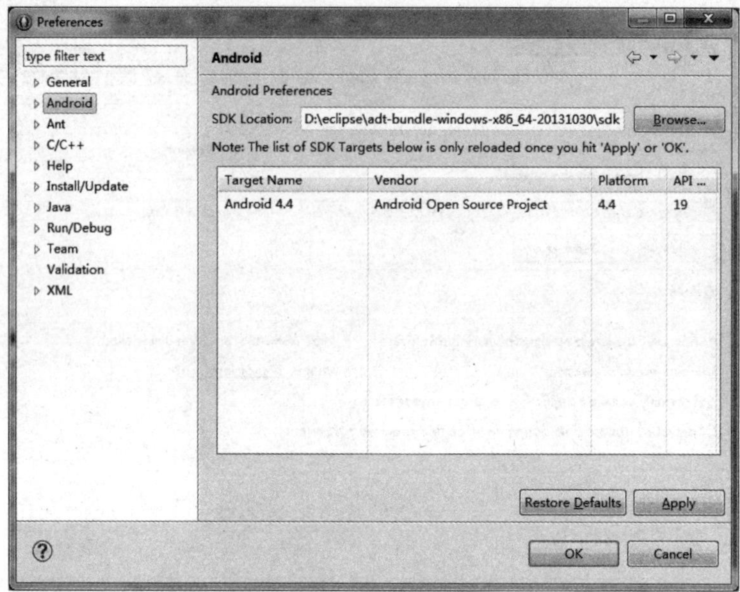

图 1.28 “首选项”对话框

技能点 3 Eclipse 项目结构

Eclipse 项目结构如图 1.29 所示。

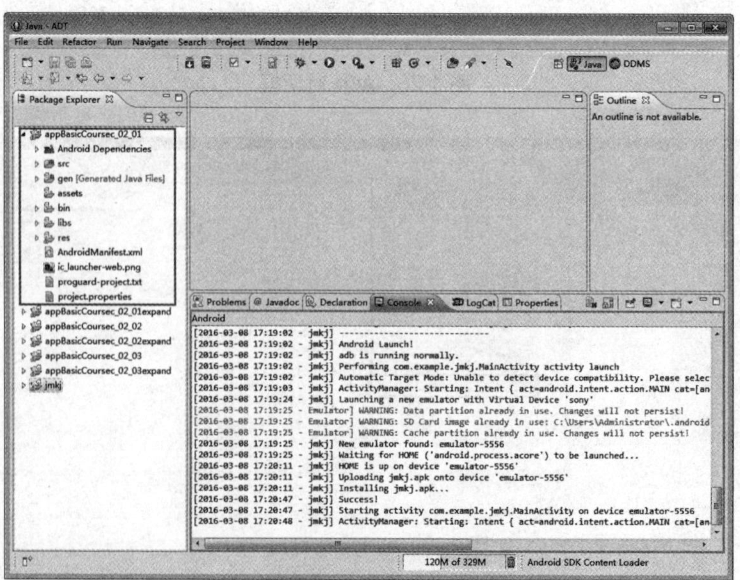

图 1.29 项目结构

（1）src 目录：主要存放 Android 项目的源文件，在 Android 项目里写的 Java 代码都在该文件下。

（2）gen 目录：自动生成的 Java 文件，里面有 2 个文件。

- BulidConfig.java：文件不需修改，成员变量 DEBUG，代表允许调试。
- R.java：文件同样不需修改，里面定义了许多静态的整型常量，是项目中使用的资源 ID。

（3）Android 4.2.2 目录：Android 开发过程中使用的 API 在该目录下 jar 包中。

（4）Android Dependencies 目录：为了兼容一些高版本的特性可以在低版本使用，不过需要注意，该 jar 包的真实位置在 libs 目录下。

（5）assets 目录：资源文件夹。但是，该文件夹里的资源并不会自动生成资源 ID 存在 R.jar 里，里面存放的是大型的资源，比如视频、音乐等。

（6）bin 目录：是二进制，是存放打包编译后的文件的，不管文件有没有编译都会在这里面。

（7）libs 目录：存放第三方库，自动被导入。

（8）res 目录：存放资源的地方，比如图片等较小的资源，有资源 ID，存放在 R.java 中。

- 在 res 中有三个名为“drawable”的文件夹，是用来存放程序所用的图片的。三个文件夹分别存放三种不同分辨率的图片，分别为“高分辨率”“低分辨率”“中分辨率”。
- 在 res 中还有一个文件夹“layout”，这个文件夹是用来存放界面布局文件的。

（9）AndroidManifest 文件：这个文件在所有的项目中名称都不会变，是一个全局的文件，所有在 Android 中用到的组件都要在该文件中声明。

技能点 4　Android 常用开发工具及用法

1　模拟器简介及功能

模拟器指在电脑上模拟 Android 系统，用模拟器来调试并运行开发的 Android 程序，通过电脑模拟器模拟手机运行环境，即可开发出应用在手机上的软件。如图 1.30 所示。

图 1.30　游戏测试

Android 模拟器的功能除了接听和拨打电话等所有移动设备上的典型功能和行为外，还可以提供大量的导航和控制键，让开发人员可以通过鼠标或键盘点击这些按键来为自己的应用程序产生事件。模拟器的屏幕用于显示 Android 自带应用程序和开发人员自己的应用程序。Android 允许开发人员的应用程序通过 Android 平台服务调用其他程序、访问网络、播放音频和视频、保存和传输数据、通知用户、渲染图像过渡和场景等。模拟器同样具有强大的调试能力，例如它能够记录内核输出的控制台、模拟程序中断（如接收短信或打入电话）、模拟数据通道中的延时效果和遗失等。

2 DDMS 调试

DDMS（Dalvik Debug Monitor Service）是 Android 开发环境中的 Dalvik 虚拟机调试监控服务。在 Android 系统平台中，每一个 Android 应用都运行在一个 Dalvik 虚拟机实例里，每一个虚拟机实例都是一个独立的进程空间。虚拟机的线程机制，内存分配和管理，Mutex 等都是依赖底层操作系统而实现的。因为 Android 应用的线程都对应一个 Linux 线程，虚拟机就可以更多地依赖操作系统的一种机制，这种机制叫线程调度和管理机制。

而 DDMS 在 IDE 与设备或模拟器之间起着调度的作用，所以它启动时会与 ADB 之间建立一个 Device Monitoring Service 用于监控设备。当设备断开或链接时，这个 Service 就会通知 DDMS 做出相应的反应。

当设备连接调试时，DDSM 和 ADB 之间会建立 VM Monitoring Service 用于监控虚拟机，并且通过 ADB Deamon 与虚拟机的 debugger 建立链接，DDMS 便可对虚拟机进行截屏、查看线程、堆的信息、LogCat 日志、进程管理、广播状态信息、模拟来电呼叫、短信和虚拟地理坐标等操作。

启动 DDMS 有三种方法。

第一种：首先选中 Eclipse 选项栏中的“Window”，然后依次选择“Open Perspective”→“DDMS”，点击启动。如图 1.31 所示。

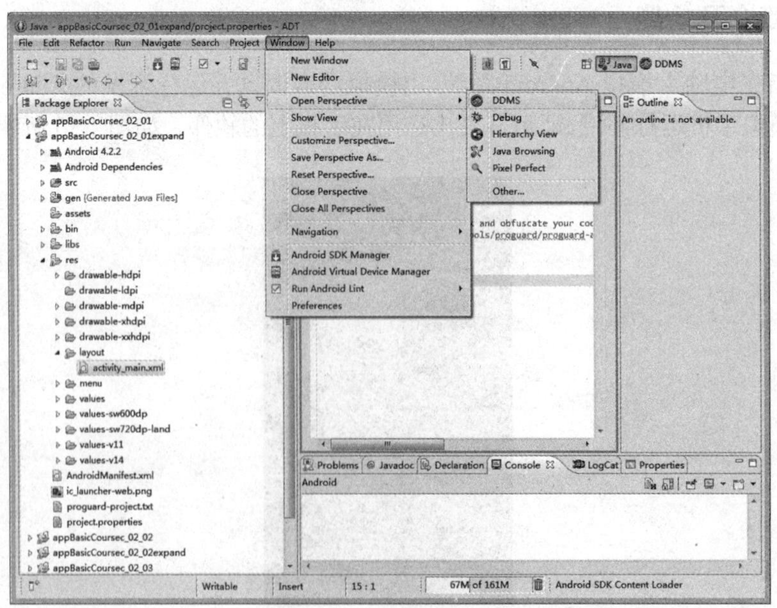

图 1.31 启动 DDMS

第二种：安装完成 ADT 后，Eclipse 上方的选项栏中会有一个 DDMS 视图。如图 1.32 所示。

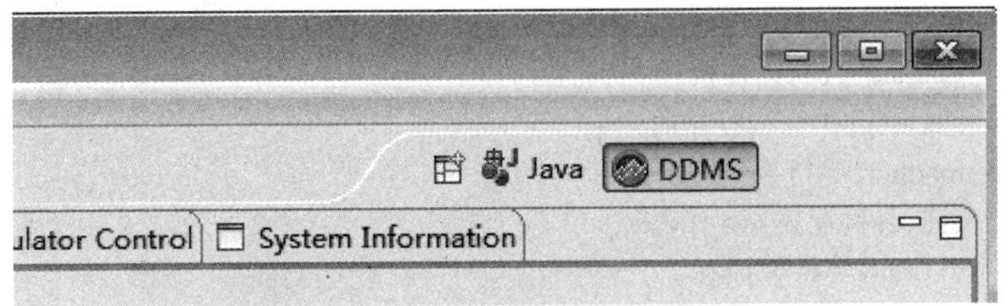

图 1.32　DDMS 的 perspective

第三种：点击如图 1.33 所示图标，选中 DDMS 选项并双击。如图 1.34 所示。

图 1.33　选项栏

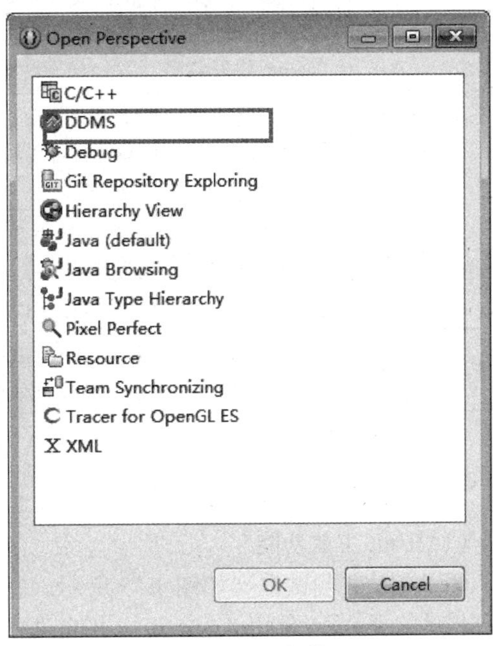

图 1.34　对话框

“DDMS”启动后界面中的各个调试窗口的作用如表 1.2 所示。

表 1.2　DDMS 调试窗口

调试窗口	描述
Devices	Device 窗口罗列模拟器中所有的进程，右上角那一排按钮分别为：调试某个进程、更新某个进程、更新进程堆栈信息、停止某个进程、最后一个图片按妞按下时抓取 android 目前的屏幕
VM Heap	当选定一个虚拟机时，VM Heap 视图不显示数据，点击右边的“Show heap updates”按钮，然后点击“Cause GC ”实施垃圾回收更新堆的状态

续表

调试窗口	描述
Threads	列出了此进程的所有线程状态 running: 代码正在执行中 sleeping: 执行线程睡眠 monitor: 等待接受监听锁 native: 执行 native 代码 vmwait: 等待虚拟机 zombie: 线程在垂死的进程 init: 线程在初始化 starting: 线程正在启动 utime: 执行用户代码的累计时间 stime: 执行系统代码的累计时间 name: 线程名字
Allocation Tracker	在这个视图里，可以跟踪每个选中的虚拟机的内存分配情况。点击“Start Tracking”后点击“Get Allocations ”就可以看到
Emulator Control	模拟一些设备状态和行为 Telephony Status: 改变电话语音和数据方案的状态，模拟不同的网络速度 Telephony Actions: 发送模拟电话呼叫和短信到模拟器 Location Controls: 发送虚拟的定位数据到模拟器，可执行定位之类的操作
File Explorer	通过 Device → File Explorer 就打开 File Explorer。这里可以浏览文件，上传下载删除文件，当然这是有相应权限限制的

3 JUnit 测试

（1）JUnit 主要功能

- Android 测试框架如图 1.35 所示。它基于 JUnit，并使用 JUnit 来测试一些与 Android 平台相关的类，或者使用 Android 的 JUint 扩展来测试 Android 组件。
- Android JUint 扩展提供了对 Android 特定组件（如 Activity，Service）的测试支持，这些扩展类提供了一些辅助方法来帮助创建测试使用的类或方法。

（2）JUnit 的好处

- 可以使测试代码与产品代码分开，有利于代码的打包和测试代码管理。
- 针对某一个类的测试代码通过较少的改动便可以应用于另一个类的测试，JUnit 提供了一个便携测试类的框架，使测试代码的编写更加方便。
- 易于集成到测试人员的构建过程中，JUnit 和 Ant 的结合可以实施增量开发。
- JUnit 是公开源代码的，可以进行二次开发。
- JUnit 具有很强的扩展性，可以方便地对 JUnit 进行扩展。

（3）JUnit 单元测试编写原则

- 简化测试的编写，这种简化包括测试框架的学习和实际测试单元的编写。
- 使测试单元保持持久性。

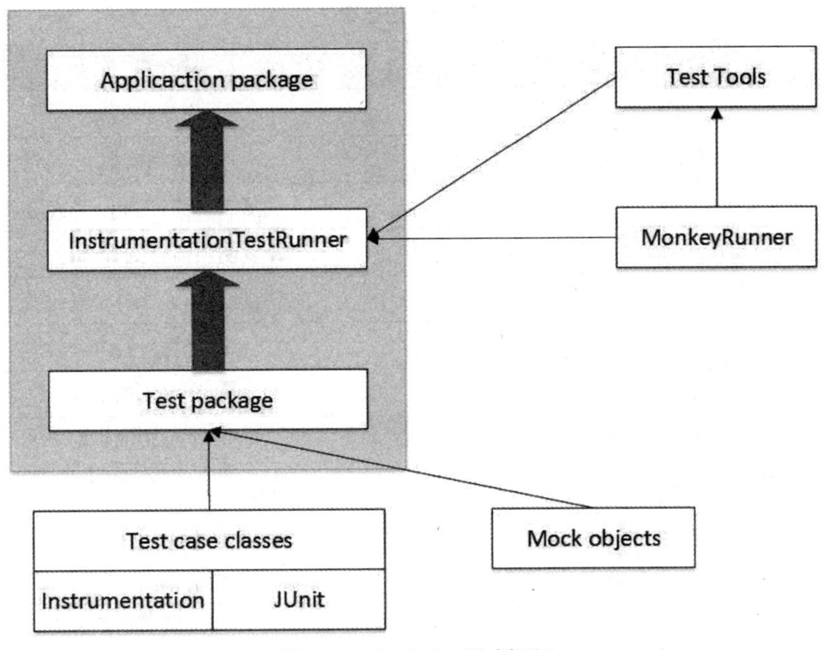

图 1.35　Android 测试框架

- 可以利用既有的测试来编写相关的测试。

(4)JUnit 的特征

- 使用断言方法判断期望值和实际值差异,返回 Boolean 值。
- 测试驱动设备使用共同的初始化变量或者实例。
- 测试包结构便于组织和集成运行。
- 支持图型交互模式和文本交互模式。

4　ADB

ADB 的全称为 Android Debug Bridge,位于 Android SDK 安装目录的"platform-tools"子目录下。利用 ADB 工具的前提是在手机上打开 USB 调试,然后通过数据线连接电脑。在电脑上使用命令模式来操作手机,可进行重启、进入 Recovery、进入 Fastboot、推送文件等功能的操作。

技能点 5　Eclipse 快捷键

在开发的时候,使用快捷键会更快更准确地编写程序,部分快捷键及其用法如表 1.3 所示。

表 1.3 快捷键

快捷键	快捷键功能
Ctrl+1	快速修复
Ctrl+D	删除当前行
Ctrl+Alt+ ↓	复制当前行到下一行
Ctrl+Alt+ ↑	复制当前行到上一行
Alt+ ↓	当前行和下面一行交互位置
Alt+ ↑	当前行和上面一行交互位置
Alt+ ←	前一个编辑的页面
Alt+ →	下一个编辑的页面
Alt+Enter	显示当前选择资源的属性
Shift+Enter	在当前行的下一行插入空行
Shift+Ctrl+Enter	在当前行插入空行
Ctrl+Q	定位到最后编辑的地方
Ctrl+M	最大化当前的 Edit 或 View
Ctrl+/	注释当前行，再按则取消注释
Ctrl+T	快速显示当前类的继承结构
Ctrl+O	快速显示 OutLine（概述）
Ctrl+L	定位在某行
Ctrl+W	关闭当前 Editer（编辑）
Ctrl+K	参照选中的 Word 快速定位到下一个
Ctrl+E	快速显示当前 Editer 的下拉列表
Ctrl+/(小键盘)	折叠当前类中的所有代码
Ctrl+×(小键盘)	展开当前类中的所有代码
Ctrl+Space	完成一些代码的插入
Ctrl+Shift+E	显示管理当前打开的所有的 View 的管理器
Ctrl+Shift+F4	关闭所有打开的 Editer
Ctrl+Shift+F	格式化当前代码
Alt+Shift+R	重命名
Alt+Shift+M	抽取本地变量
Ctrl+Z	撤销当前的操作
Alt+Shift+C	修改函数结构

第一步：打开 Eclipse，新建一个工程文件。步骤如图 1.36 所示。

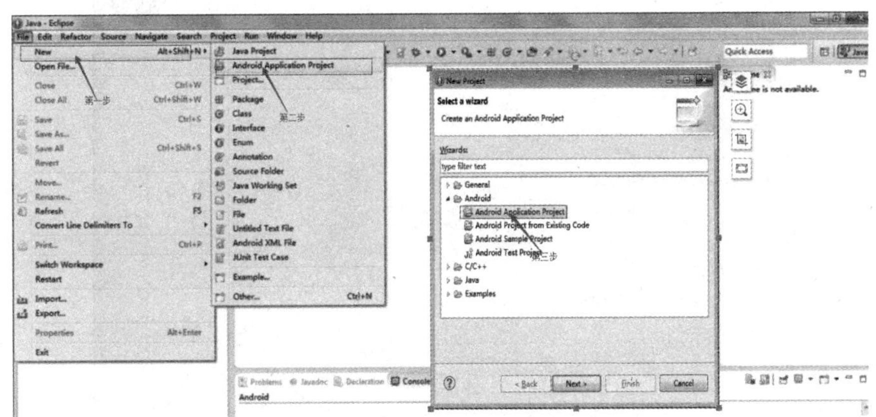

图 1.36　创建新的工程文件

第二步：应用程序名称与工程名称为"HelloWorld"，包名自动生成。选择 SDK 的版本，即程序最低支持的 Android 版本。步骤如图 1.37 所示。

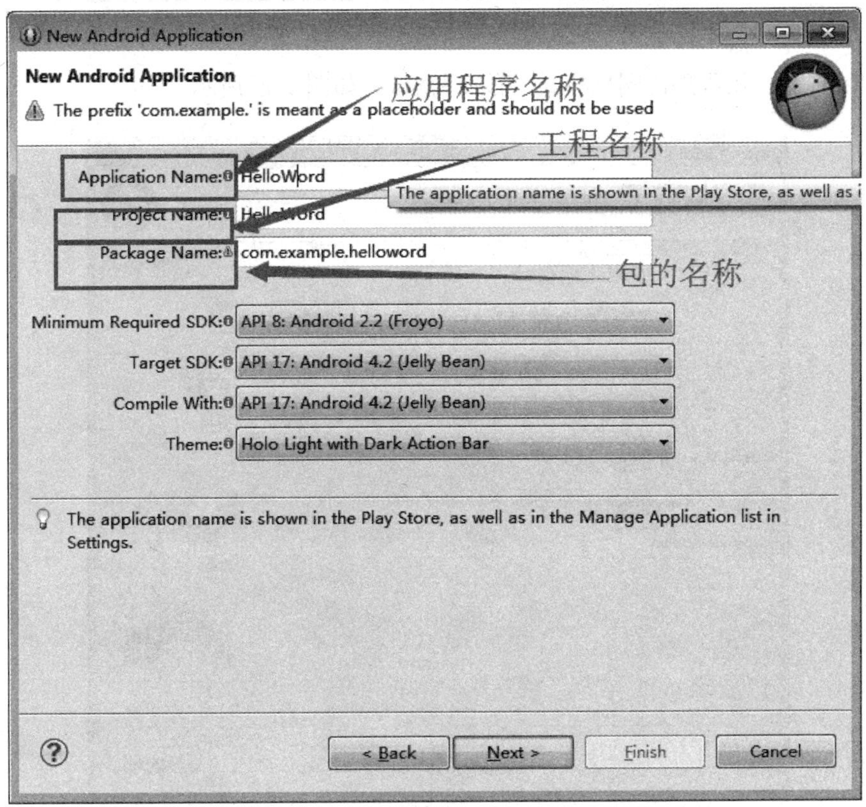

图 1.37　工程信息设置

第三步：创建登录图标（launcher icon）和 activity。默认选中，选择工作空间（Workspace）。步骤如图 1.38 所示。

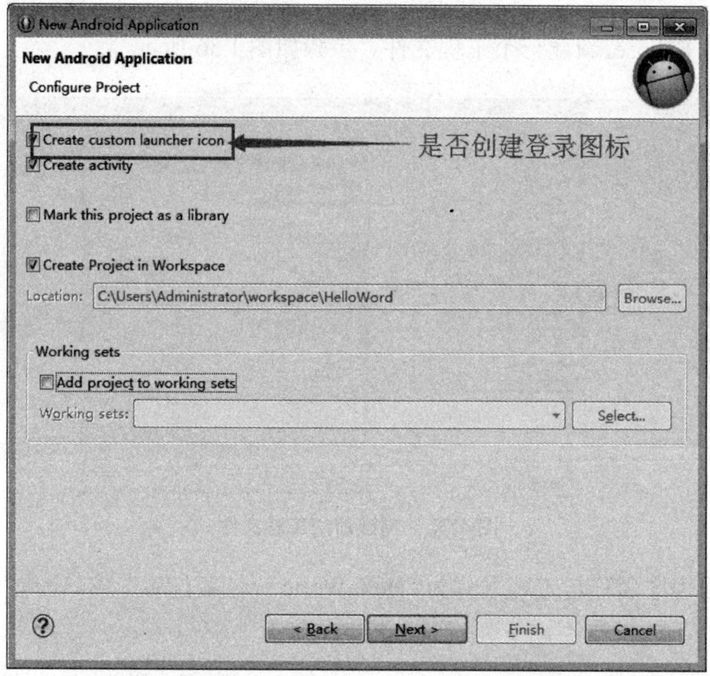

图 1.38 New Android Application 界面

第四步：选择应用程序的图标样式，可选择默认。如图 1.39 所示。

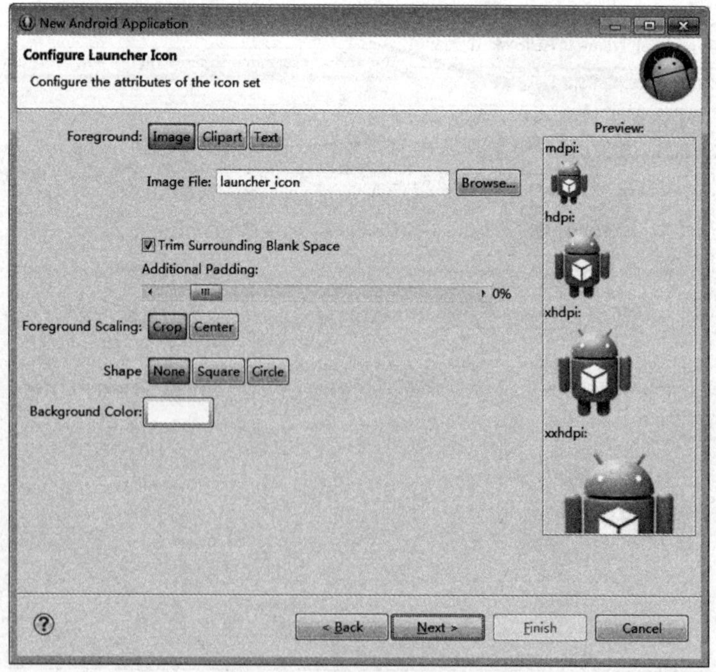

图 1.39 图标样式选择框

第五步：点击“Finish”，完成创建。创建完成项目之后 Eclipse 自动生成代码。如图 1.40 所示。

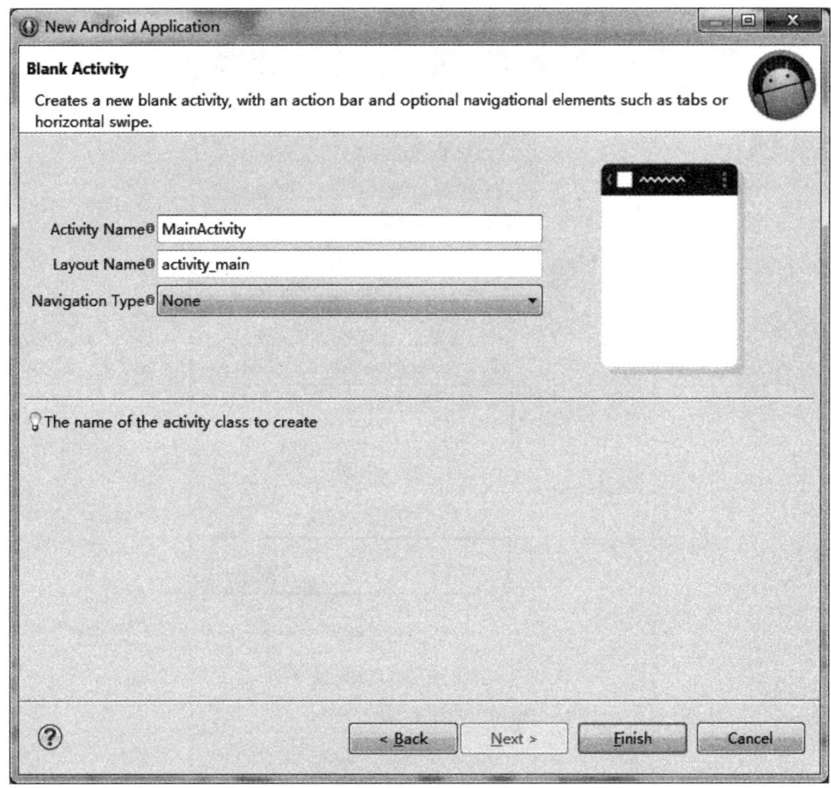

图 1.40　Activity Name

第六步：出现该图标时，证明环境已搭建完成，单击左上方的第二个机器人。如图 1.41 所示。

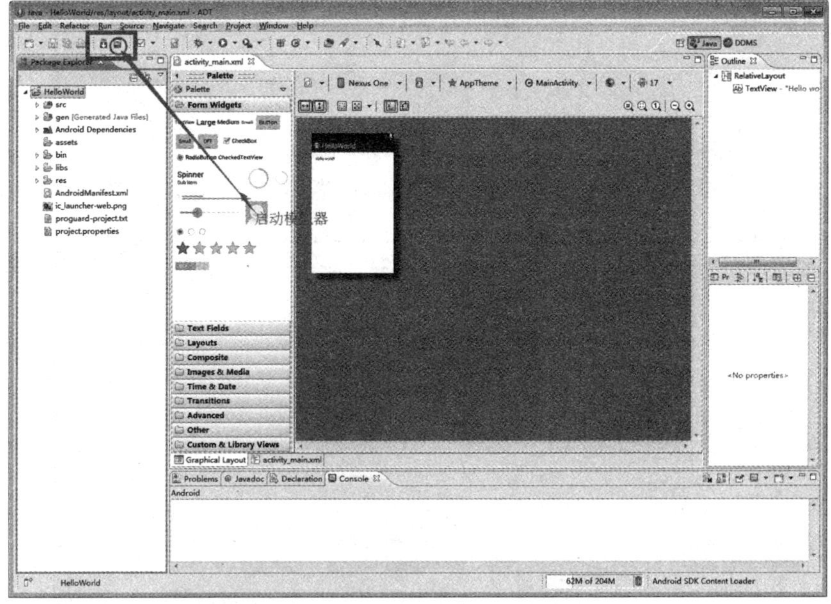

图 1.41　Eclipse 运行编程界面

第七步：在 AVD name 栏根据自己的项目设定名称。根据自己的需要可以自行分配内存卡大小。如图 1.42 所示。

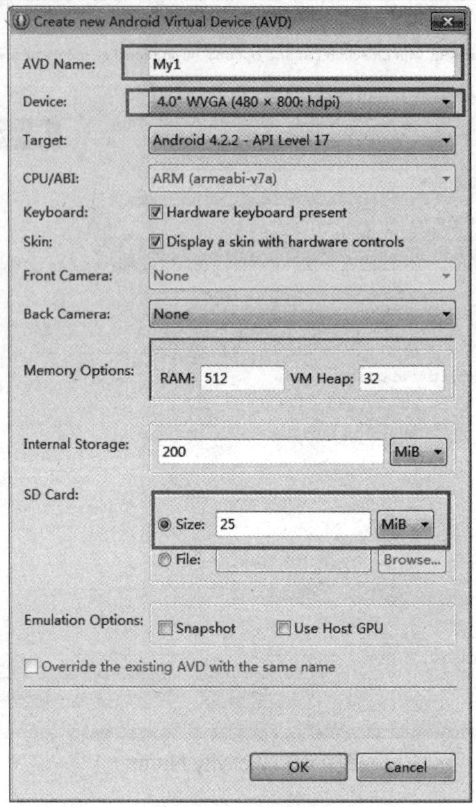

图 1.42 模拟器创建界面

第八步：选中创建的模拟器，单击“Start...”选项。如图 1.43 所示。

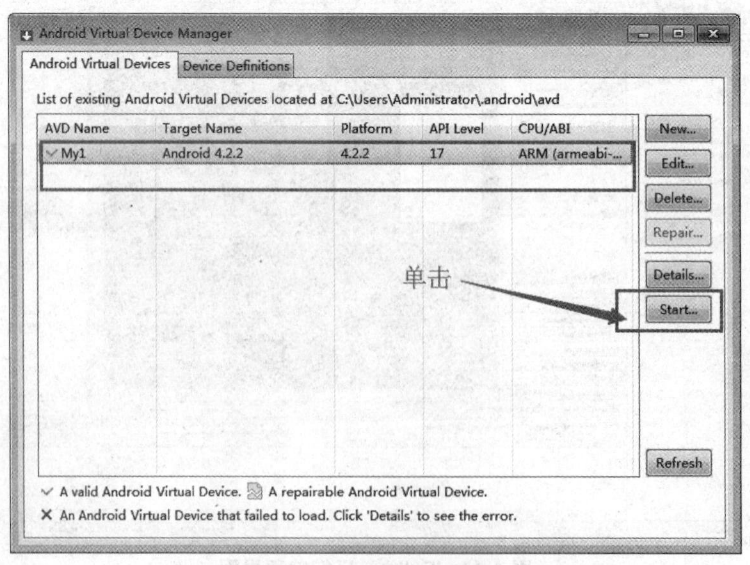

图 1.43 模拟器选项框

第九步：选择“Launch”，启动模拟器。如图 1.44 所示。启动后得到界面如图 1.45 所示。

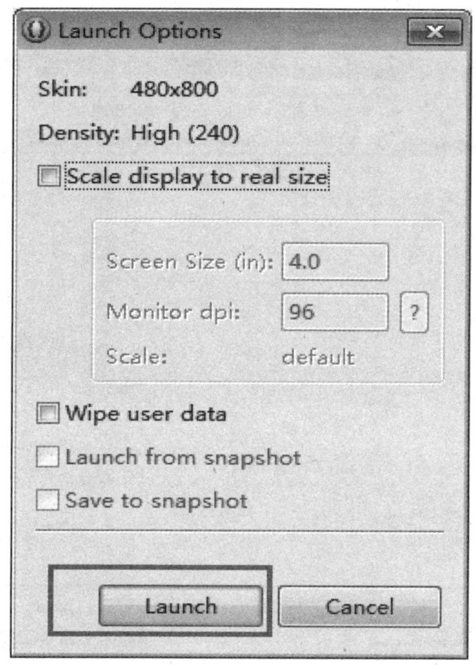

图 1.44　模拟器启动界面

图 1.45　模拟器界面

第十步：选中 HelloWorld 文件右击，依次选择：“Run As”→“Android Application”，等待模拟器运行，得到最后的结果。如图 1.46 所示。

图 1.46　HelloWrod 应用程序

第十一步：选择“Window”→“Open Perspective”→“DDMS”，启动“DDMS”，得到如图 1.47 所示的界面。

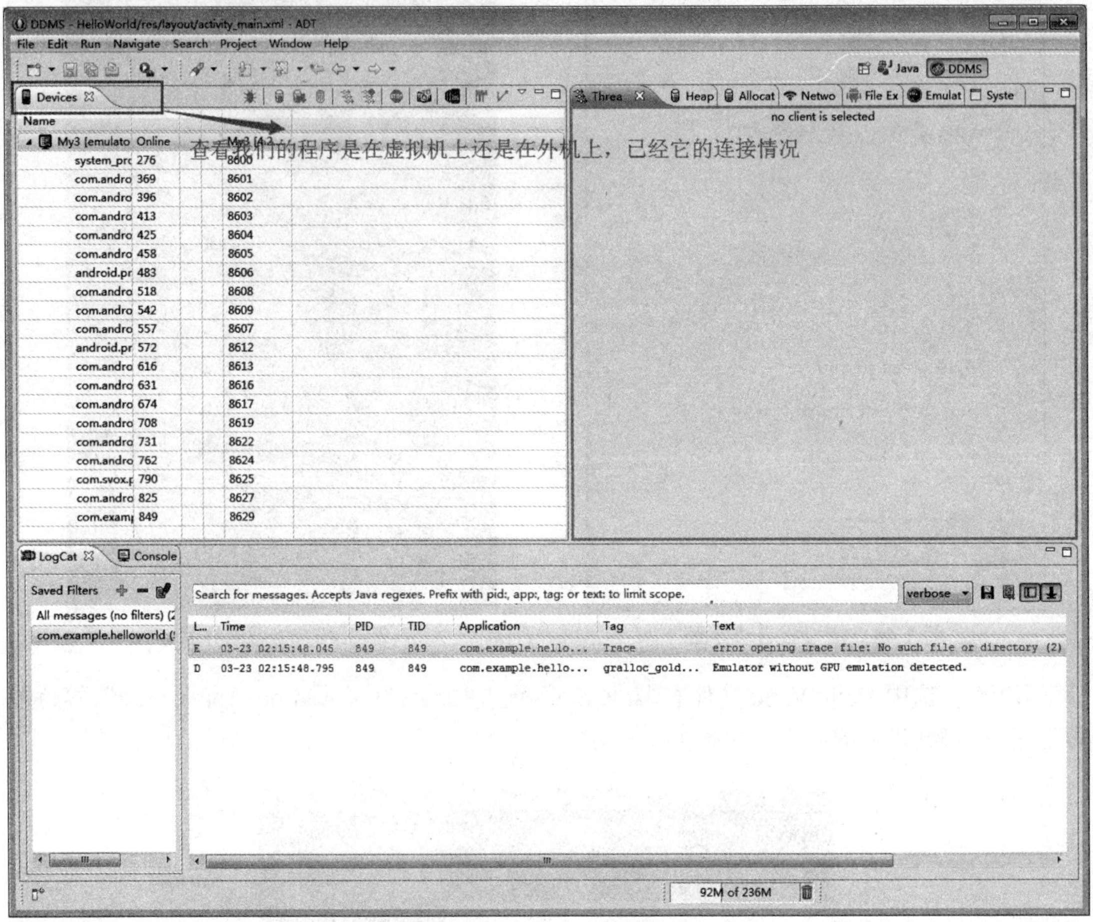

图 1.47 DDMS 界面

本项目介绍了 Android 手机平台的基础知识，重点讲解如何搭建和使用 Android 系统平台。通过对本项目的学习，可以清楚地了解 Android 开发的基本概念，掌握 Eclipse 的特点、环境的搭建和程序运行的基本流程，提高对 Android 开发的认知度。

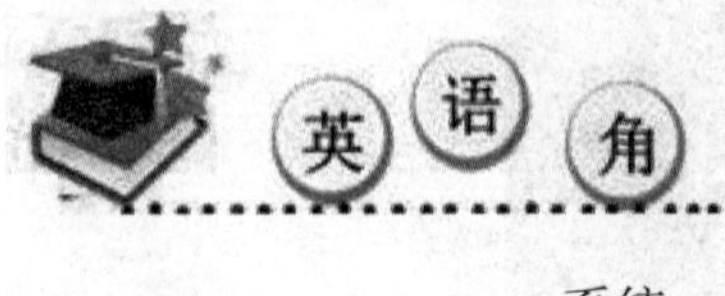

system	系统
manager	处理者

project　　项目
logo　　标志
Android　　安卓
font　　文字
install　　设置
window　　窗口
tool　　工具
finish　　结束

一、选择题

1. 下列哪一项不是 Android 的功能？（　　）
A. 存储　　B. 消息传递　　C. 媒体支持　　D. 连接桥梁
2. 下列哪一项不是 Android 的架构层？（　　）
A. 应用程序层　　B.Linux 内核层　　C. 库　　D.dao 层
3. 下列哪一项不是 Android 的优势？（　　）
A. 开放性　　B. 多样硬件设施
C. 与 Google 的无缝结合　　D. 操作系统的秘密性
4. 下列哪一项是 Android 开发不需要的？（　　）
A.SDK　　B.ADT　　C.MyEclipse　　D.JDK
5. Android 用来存储数据的数据库是（　　）。
A. TomCat　　B. Sqlite　　C.Oracle　　D.MySql

二、填空题

1. _______ 的全称为 Android Debug Bridge，起到调试桥的作用。
2. 使用 ADB 我们可以在电脑上对手机进行：________、___________、___________、_____________ 等功能的操作。
3. _____________ 主要是对程序员开放的 Android 操作系统的各种功能，以便在应用程序中使用各项功能。
4. _____________ 主要包含一些 C/C++ 库，这些库能被 Android 系统中不同的组件使用。
5. _____________ 为 Android 的内核，包括 Android 设备的各种硬件组建的底层设备驱动程序。

三、判断题

1. ADT 是对 Eclipse IDE 的扩展，用来支持 Android 应用程序的创建和调试。（　　）
2. Android 是基于 C 并运行在 Linux 内核上的轻量级操作系统。（　　）

3. 从架构图看，Android 分为三个层。 ( )

4. Android 是一种语言。 ( )

5. 安装 JDK 时候只能安装到默认路径。 ( )

四、简答题

1. 什么是 ADB，它有什么作用？

2. 描述一下 Android 框架。

五、上机题

根据“HelloWorld”应用程序开发过程，开发一个名为“How are you in China”的应用程序。

项目二　Android 应用界面设计

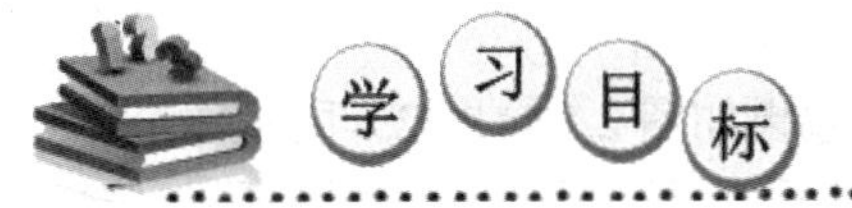

通过实现应用界面的相关设计，学习常用控件和 Dialog 的相关知识，了解视图与布局相互结合以及 Dialog 组件的使用。在项目实现过程中：

- 了解视图与 View 组件；
- 掌握 Android 应用界面的设计布局技术；
- 掌握 Dialog 组件相关的知识和技能。

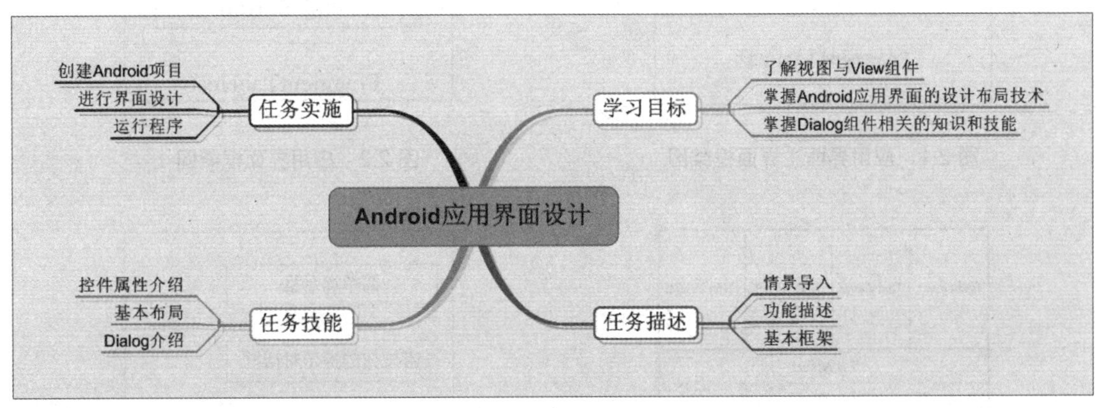

【情景导入】

Android 市场不断拓展，用户对 Android 界面设计的要求也越来越高，希望有一个美观、整洁、实用的界面，因此 Android 提供了一些常用控件来实现界面的设计。本次任务主要是设计应用界面。

【功能描述】

本项目将使用其中典型的控件来设计 UI 界面。

- 使用线性布局技术来设计应用界面主界面布局；
- 使用 FragmentTabHost 实现三个界面切换；
- 挑选界面控件栏中控件，使用线性布局设计第一个界面；
- 使用常用布局设计第二个界面；
- 在第三个界面实现对话框功能。

【基本框架】

基本框架如图 2.1 至图 2.4 所示，将框架图转换成的效果如图 2.5 至图 2.8 所示。

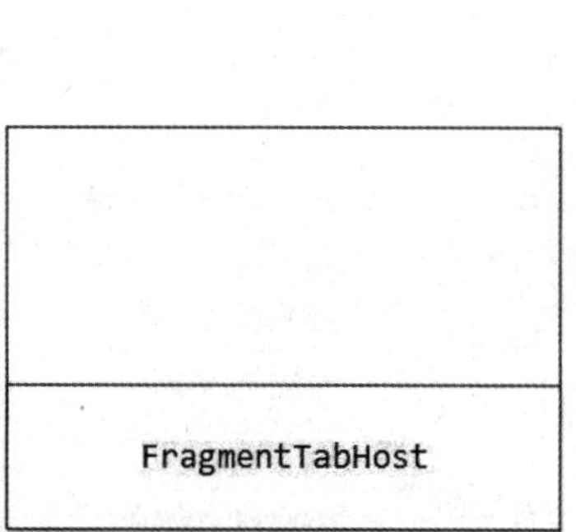

图 2.1 应用界面主界面框架图

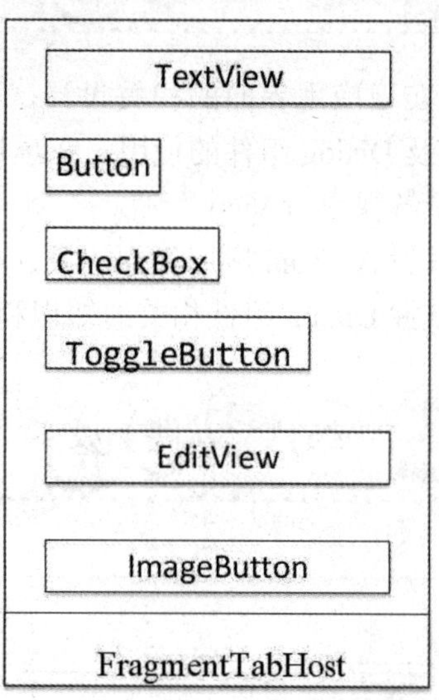

图 2.2 应用界面框架图 1

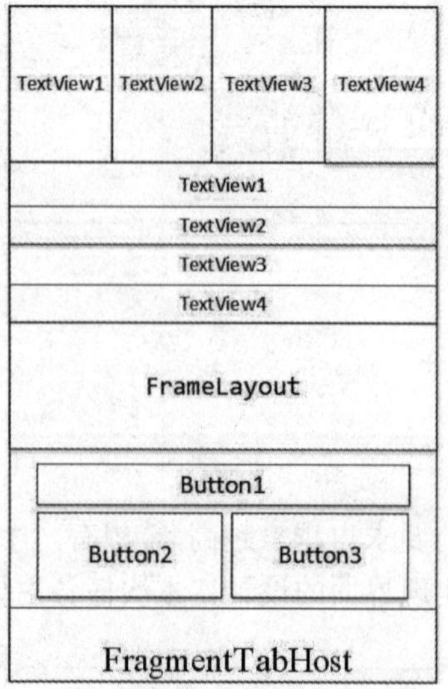

图 2.3 应用界面框架图 2

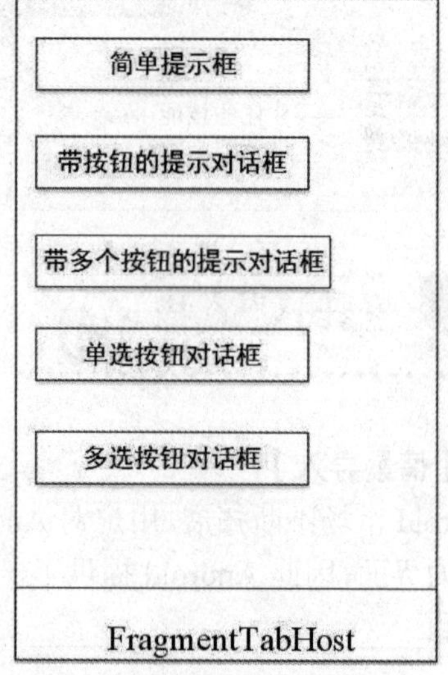

图 2.4 应用界面框架图 3

图 2.5　应用界面主界面效果图

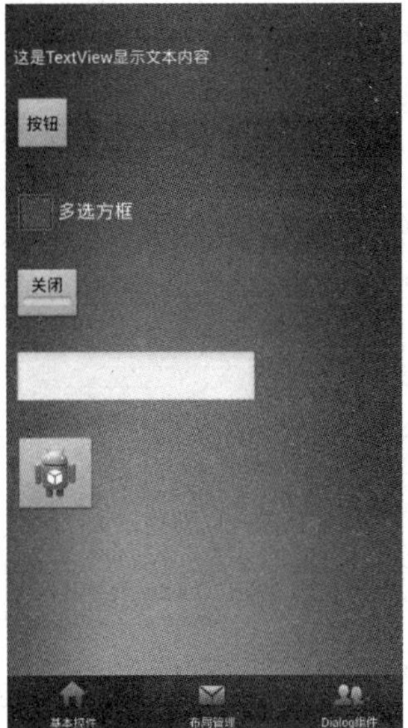

图 2.6　应用界面效果图 1

图 2.7　应用界面效果图 2

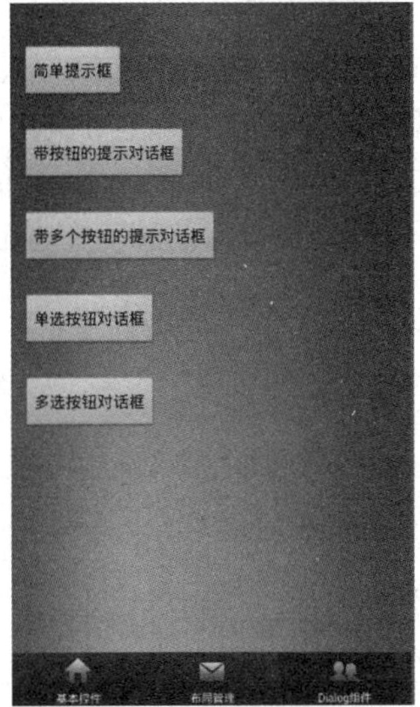

图 2.8　应用界面效果图 3

技能点 1　控件属性介绍

各个控件的继承关系如图 2.9 所示。了解各个控件的继承关系对之后项目的内容理解有很大的帮助。

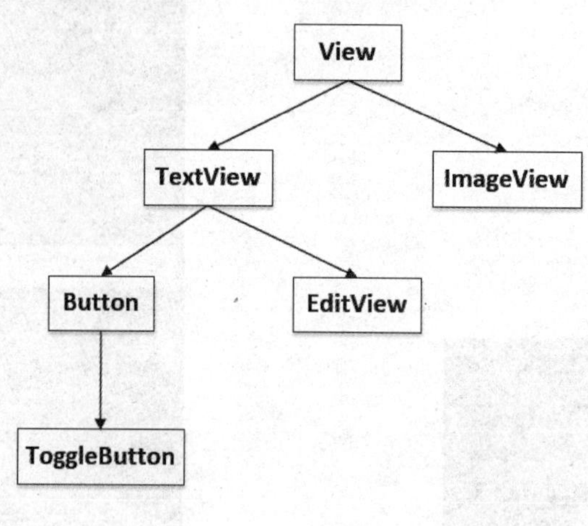

图 2.9　控件继承关系图

1　文本框与编辑框

TextView 继承了 View，是 EditView 的父类。TextView 是一个文本编辑器，但 Android 关闭了它的编辑功能。开发一个可编辑的文本框可使用它的子类 EditView，EditView 是允许用户编辑的文本框。EditText 与 TextView 共用大部分 XML 属性和方法，二者之间的区别为 EditText 可接受用户输入信息。文本框与编辑框属性及描述如表 2.1 所示。

表 2.1　文本框与编辑框属性及描述

XML 属性	描述
android:layout_width	设置组件的宽度
android:layout_height	设置组件的高度
android:id	给组件定义一个 id 值，供后期使用
android:background	设置组件的背景颜色或背景图片
android:text	设置组件的显示文字
android:textColor	设置组件的显示文字的颜色
android:layout_below	组件在参考组件的下面
android:alignTop	同指定组件的顶平行
android:maxLength=" "	限制输入字数
android:password='true'	可以让 EditText 显示的内容自动为星号
android:numeric='true'	true：让输入法自动变为数字输入键盘，同时仅允许 0~9 的数字输入
android:textColor	显示文本颜色
android:textSize	设置文字大小，推荐度量单位“sp”，如“15sp”

使用 TextView 属性实现如图 2.10 所示效果，代码如下：

```
<TextView
        android:id="@+id/textView1"// 设置 TextView 控件 id
        android:layout_width="wrap_content"// 设置 TextView 控件宽度
        android:layout_height="wrap_content"// 设置 TextView 控件高度
        android:layout_marginTop="30dp"// 设置 TextView 控件距顶层距离
        android:text=" 这是 TextView 显示文本内容 "// 添加 TextView 控件文本
        android:textColor="#fff" />// 添加 TextView 控件文本颜色为白色
```

图 2.10　TextView 效果图

2　Button 组件

Button 继承了 TextView，可供用户点击，点击的时候时会触发 onClick 事件（点击事件）。按钮使用简单，可设置 Button 的背景以及文字等属性，如果背景为不规则图片则能够开发出不同规则形状的按钮。Button 的属性设置方法及描述如表 2.2 所示。

表 2.2　Button 属性设置方法及描述

方法	描述
setClickable(boolean clickable)	clickable=true: 允许点击 clickable=false: 禁止点击
setBackgroundResource(int resid)	通过资源文件设置背景色 resid: 资源 xml 文件 ID
setText(CharSequence text)	设置按钮显示文字
setTextColor(int color)	设置按钮显示文字的颜色 color 可以使用系统 Color 常量
setOnClickListener(OnClickListener l)	设置按钮点击事件

使用 Button 属性实现如图 2.11 所示效果，代码如下：

```
<Button
        android:id="@+id/button1"
        android:layout_width="wrap_content"
        android:layout_height="wrap_content"
```

```
        android:layout_marginTop="24dp"
        android:text=" 按钮 " />
```

图 2.11　Button 效果图

3　状态开关

ToggleButton 按钮是由 Button 派生出来的,所以 Button 的所有属性、方法也同样可以用于 ToggleButton。它的作用一般是切换程序中的某种状态。ToggleButton 属性设置方法及描述如表 2.3 所示。

表 2.3　ToggleButton 属性设置方法及描述

XML 属性	描述
android:checked	设置该按钮是否被选中
android:textOff	设置当该按钮的状态关闭时显示的文本
android:textOn	设置当该按钮的状态打开时显示的文本

使用 ToggleButten 属性实现如图 2.12 所示效果,代码如下:

```
<ToggleButton
        android:id="@+id/toggleButton1"
        android:layout_width="wrap_content"
        android:layout_height="wrap_content"
        android:layout_marginTop="24dp"`
        android:textOff=" 关闭 "
        android:textOn=" 开启 "
        />
```

图 2.12　ToggleButton 效果图

4　ImageView 及其子类

ImageView 继承自 View 组件，其功能是显示图片，可使用 ImageView 显示 Drawable 对象。ImageView 常用方法及说明如表 2.4 所示。

表 2.4　ImageView 常用方法及说明

方法	描述
setAlpha(int alpha)	设置 ImageView 的透明度
setImageBitmap(Bitmap bm)	设置 ImageView 所显示的内容为指定的 Bitmap 对象
setImageDrawable(Drawable drawable)	设置 ImageView 所显示的内容为指定的 Drawable 对象
setImageResource(int resId)	设置 ImageView 所显示的内容为指定 Id 的资源
setImageURI(Uri uri)	设置 ImageView 所显示的内容为指定 Uri
setSelected(boolean selected)	设置 ImageView 的选中状态

使用 ImageView 属性实现如图 2.13 所示效果，代码如下：

```
<ImageView
        android:id="@+id/imageview"
        android:layout_width="wrap_content"
        android:layout_height="wrap_content"
        android:focusable="false"// 设置 ImageView 的焦点
        android:padding="3dp" // 设置 ImageView 距外边框的距离
        android:src="@drawable/tab_home_btn">
</ImageView>
```

图 2.13　ImageView 效果图

5　列表视图与网格视图

ListView 是手机系统中使用非常广泛的一种组件，它继承自 LinearLayout 布局，以垂直（vertical）列表的形式显示所有列表项。LinearLayout 布局会在技能点二中进行讲解，生成列表视图有如下两种方式。

- 直接使用 ListView 进行创建；
- 让 Activity 继承 ListActivity。

GridView 与 ListView 相似性极高，都是列表项，但是 ListView 只可以显示一列，而 GirdView 可显示多列。ListView 与 GridView 都会使用到适配器。常用适配器如表 2.5 所示。

表 2.5 常用适配器

适配器	描述
ArrayAdapter<T>	用来绑定一个数组，支持泛型操作
SimpleAdapter	用来绑定在 XML 中定义的控件对应的数据
SimpleCursorAdapter	用来绑定游标得到的数据
BaseAdapter	通用的基础适配器

拓展：想了解或学习 Android 更多的控件属性知识，可扫描下方二维码，获取更多信息。

技能点 2 基本布局

不同型号的手机屏幕分辨率和尺寸不完全相同，如果让程序手动控制每个组件的大小位置，会给编程带来很大的困难。为了让不同组件在不同的手机屏幕上都能运行良好，Android 提供了布局管理器，可以让程序员根据运行平台调整组件的大小，选择合适的布局管理器。在 Android 系统中提供了五种基本的布局方式，通过这五种布局方式，能够实现大多数复杂界面的设计。继承关系如图 2.14 所示。

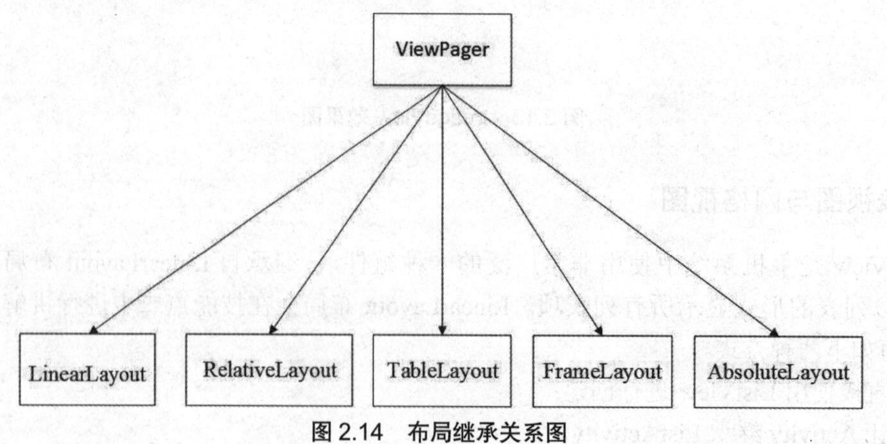

图 2.14 布局继承关系图

1　线性布局 (LinearLayout)

线性布局是将控件排放在水平方向或者垂直方向的一条线上。在线性布局中还有一些比较常用的属性与设置方法，如表 2.6 所示。

表 2.6　线性布局的常用属性

XML 属性	对应设置方法	描述
android:orientation	setOrientation(int)	控制排列方式的属性 horizontal（水平排列）、vertical（垂直、默认值）
android:gravity	setGravity(int)	设置组件的对齐方式，该属性的值有 top、botton、left、right、center_vertical、fill_horizontal、center、fill、clip_vertical、clip_horizontal。也可同时使用多种对齐方式，让其组合在一起
android:divider	setDivider-Drawable(Drawable)	设置垂直布局时两个按钮之间的分隔线
android:baselineAligned	setBaselineAligned(boolean)	boolean =false，就会阻止该布局与它的子元素的基线对齐；boolean = true 则可以
android:measureWith-largestChild	setMeasureWithLargest-ChildEnabled(boolean)	boolean =true，带权重的子元素都会被设置为有最大子元素的最小尺寸

使用 LinearLayout 属性实现如图 2.15 所示效果，代码如下：

```
    <LinearLayout
        android:layout_width="match_parent"
        android:layout_height="0dp"
        android:layout_weight="1"
        android:orientation="horizontal" // 控制排列方式为竖排
  >
  </LinearLayout>
```

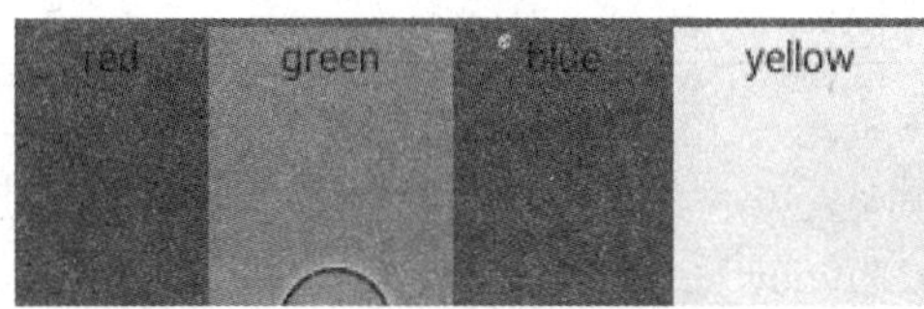

图 2.15　线性布局效果图

2　相对布局 (RelativeLayout)

相对布局是按照控件间的相对位置进行布局，可以选择某个控件作为参照，其他的控件可以在它的任意方向，例如“android:layout_below”属性指在某个控件的下面。

RelativeLayout 的主要属性如表 2.7 所示。

表 2.7 RelativeLayout 属性

XML 属性	描述
android:layout_centerHorizontal	该子组件是否在布局中水平居中
android:layout_centerVertical	该子组件是否在布局中垂直居中
android:layout_centerInparent	该子组件是否在布局中的中央位置
android:layout_below	在某元素的下方
android:layout_above	在某元素的上方
android:layout_marginBottom	离某元素底边缘的距离
android:layout_marginLeft	离某元素左边缘的距离
android:layout_alignParentRight	该子组件是否与整个布局右边对齐
android:layout_alignParentTop	该子组件是否与整个布局顶端对齐

3 表格布局 (TableLayout)

表格布局与平时使用的 Excel 表格类似，都是将子元素的位置设置到具体的行或列中。一个 TableLayout 是由许多的 TableRow 组成的，一个 TableRow 就相当于 TableLayout 中的一行。TableRow 是 LinearLayout 的一个子类，它的某些属性是一直不变的，比如说排列方式属性值恒为 horizontal，它的 android:layout_width 和 android:layout_height 属性值恒为 match_parent 和 wrap_content。因此其子元素都是横向排列并且宽高一致的。因为这些系统已经设定这样的属性，使得每个 TableRow 中的子元素都可以认为是表格中的单元格。在 TableLayout 里，单元格可以为空，但注意不能跨行。

表格布局管理器中，可为单元格设置如下三种行为方式。

Shrinkable：如果某列被设为 Shrinkable，则该列所有单元格宽度可被变大或者变小，这样将列设置为 Shrinkable 就可以让表格适应父容器的宽度。

Stretchable：如果某列被设为 Stretchable，那么该列的所有单元格宽度可被拉长或者拉短，这样就能保证组件能填满表格的剩余空间。

Collapsed：如果某列被设为 Collapsed，那么该列的所有单元格都不显示了。

使用 TableLayout 属性实现如图 2.16 所示效果，代码如下。

```
<TableLayout
        android:layout_width="match_parent"
        android:layout_height="match_parent"
        android:background="#99cc33"// 设置背景
        android:padding="4dp"
        android:shrinkColumns="0,1,2" // 拉伸所有列
>
</TableLayout>
```

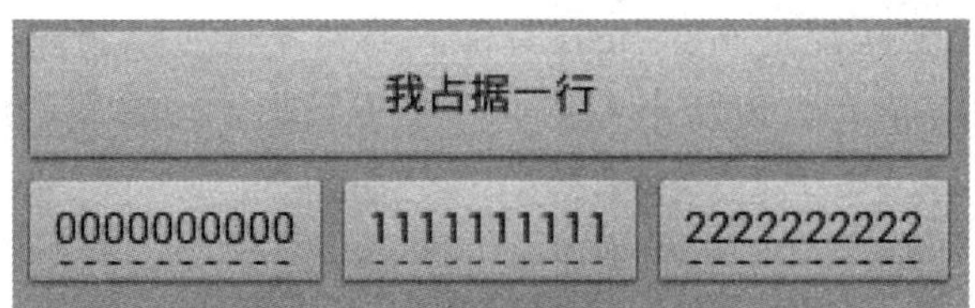

图 2.16　表格布局效果图

4　帧布局 (FrameLayout)

FrameLayout 是五大布局中最简单的一种布局，在该布局中，整个界面被当成一块空白的没有任何组件的区域，所有的子元素都不能被指定放置的位置，这一点是与其他布局的最大差别。所有的子组件系统会被默认放在这块区域的左上角，并且后面的子元素直接覆盖在前面的子元素之上，将前面的子元素部分或全部遮挡。这就像在画板上刷颜料，刷一层就会将原本位置的颜色覆盖。

FrameLayout 的大小由子控件中最大的子控件决定，如果组件都一样大的话，同一时刻只能看到最上面的组件，也可以为组件添加 layout_gravity 属性，从而制定组件的对齐方式。

5　绝对布局 (AbsoluteLayout)

相较于其他的几种布局，绝对布局是很好理解的一种布局。平时在生活中，我们的家具都是一般摆放在一个固定位置的，以房子的一个角作为坐标原点，然后这个位置就可以用 X,Y 来表示。绝对布局中通过 android:layout_x 和 android:layout_y 来指定其子元素的准确的坐标位置。当使用 AbsoluteLayout 作为布局容器时，布局容器不再管理子组件的位置、大小，这些需要程序员自行控制。

开发每种布局效果都要找到适合的布局方式，例如平时用的手机计算器，它最合适的布局就是 TableLayout（表格布局）。另外，这五个布局元素可以相互嵌套应用，做出理想中的效果。

技能点 3　Dialog 介绍

1　Dialog 简介

Dialog 是 Android 开发过程中最常用到的组件之一，它可以用来弹出一个窗体，这个窗体的内容大多用来提示或警告用户。Dialog 对话框可以分为 5 大类，分别是警告对话框（AlertDialog）、进度对话框（ProgressDialog）、日期选择对话框（DatePickerDialog）、时间选择对话框（TimePickerDialog）、自定义对话框（从 Dialog 继承）。

Dialog 的创建方式有两种：第一种是新建一个 Dialog 对象，调用 Dialog 对象的 show() 和 dismiss() 方法控制对话框的显示和隐藏。第二种在 Activity 的 onCreateDialog(int id) 方法中创建 Dialog 对象并返回，调用 Activty 的 showDialog(int id) 和 dismissDialog(int id) 来显示和隐藏对话框。其区别在于通过第二种方式创建的对话框会继承 Activity 的属性，如获得 Activity 的 menu 事件等。

AlertDialog 的构造方法都是 Protected（有保护的），不能直接新建 AlertDialog，需要使用 AlertDialog.Bulider 中的 create() 方法来创建一个弹窗窗口。想要使用 AlertDialog.Bulider 创建对话框，需要掌握以下几个方法，如表 2.8 所示。

表 2.8 设置 Dialog 窗口方法

设置方法	描述
setTitle()	为对话框设置标题
setIcon()	为对话框设置图标
setMessage()	为对话框设置内容
setView()	给对话框设置自定义样式
setItems()	设置对话框要显示的一个 list，一般用于显示几个命令
setNeutralButton()	普通按钮
setPositiveButton()	对话框添加“Yes”按钮
setNegativeButton()	对话框添加“No”按钮
Create()	创建对话框
show()	显示对话框

2 Dialog 实现步骤

（1）使用 AlertDialog.Bulider 类创建一个选择提示框。

```
AlertDialog.Builder dialog03 = new AlertDialog.Builder(activity);// 创建 Dialog 窗口
```

（2）设置提示框的各个参数并显示窗口。

```
dialog03.setTitle(" 带多个按钮的提示对话框 ");// 设置 Dialog 标题
        dialog03.setMessage(" 提示信息 ");// 添加副标题
        dialog03.setPositiveButton(" 确定 ", null); // 添加确定按钮
        dialog03.setNeutralButton(" 取消 ", null); // 添加取消按钮
        dialog03.setNegativeButton(" 删除 ", null); // 添加删除按钮
        dialog03.create().show();
```

（3）效果如图 2.17 所示。

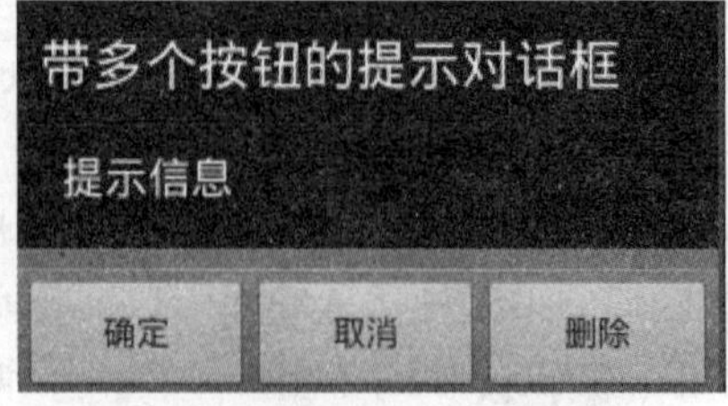

图 2.17 Dialog 对话框效果图

第一步：在 Eclipse 中创建一个 Android 工程，从选择控件栏（如图 2.18 所示）中选择所需控件设计主界面，如图 2.5 所示。具体如代码 CORE0201 所示。

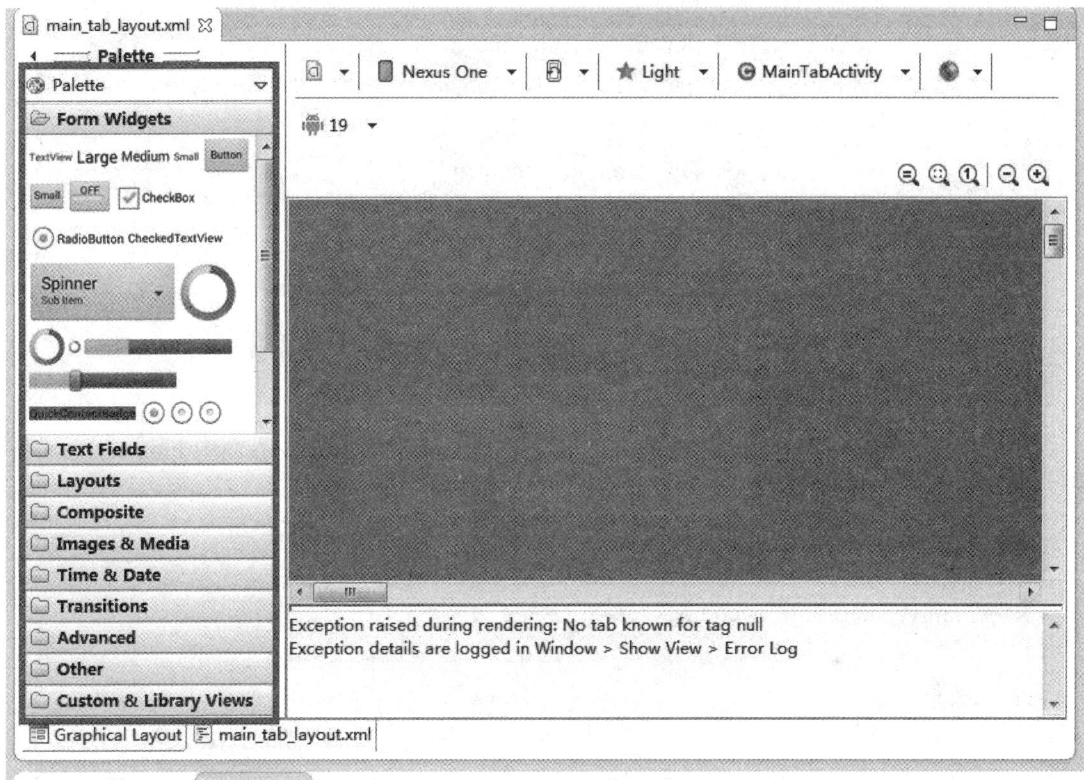

图 2.18　界面布局

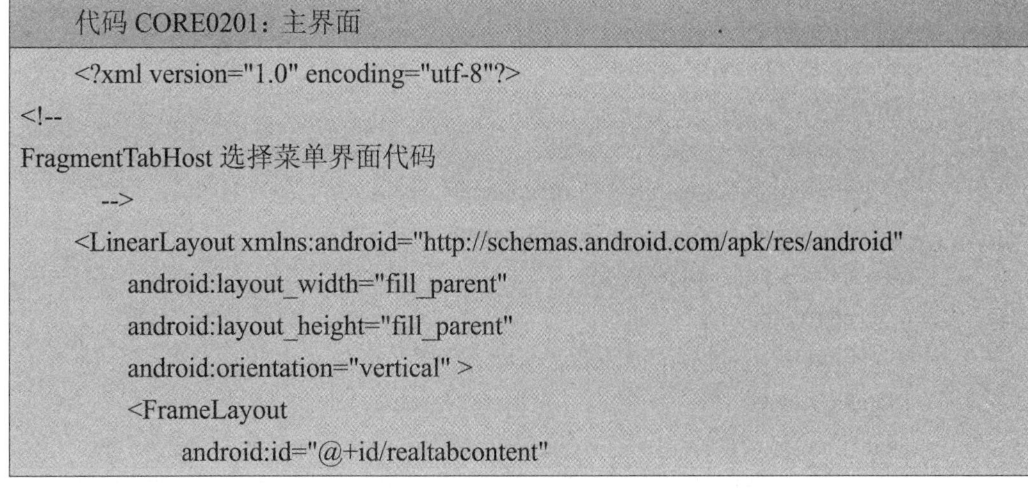

代码 CORE0201：主界面

```
<?xml version="1.0" encoding="utf-8"?>
<!--
FragmentTabHost 选择菜单界面代码
    -->
<LinearLayout xmlns:android="http://schemas.android.com/apk/res/android"
    android:layout_width="fill_parent"
    android:layout_height="fill_parent"
    android:orientation="vertical" >
    <FrameLayout
        android:id="@+id/realtabcontent"
```

```
            android:layout_width="fill_parent"
            android:layout_height="0dip"
            android:layout_weight="1" />
    <android.support.v4.app.FragmentTabHost
            android:id="@android:id/tabhost"
            android:layout_width="fill_parent"
            android:layout_height="wrap_content"
            android:background="@drawable/maintab_toolbar_bg">
            <FrameLayout
                android:id="@android:id/tabcontent"
                android:layout_width="0dp"
                android:layout_height="0dp"
                android:layout_weight="0" />
        </android.support.v4.app.FragmentTabHost>
    </LinearLayout>
```

第二步：在“res”→“layout”下创建 tab_item_view.xml 文件填充主界面中 FrameLayout 信息。具体如代码 CORE0202 所示。

代码 CORE0202：填充主界面 FrameLayout 信息

```
    <?xml version="1.0" encoding="utf-8"?>
<!--
item 界面代码
    -->
    <LinearLayout xmlns:android="http://schemas.android.com/apk/res/android"
        android:layout_width="wrap_content"
        android:layout_height="wrap_content"
        android:gravity="center"
        android:orientation="vertical" >
        <ImageView
            android:id="@+id/imageview"
            android:layout_width="wrap_content"
            android:layout_height="wrap_content"
            android:focusable="false"
            android:padding="3dp"
            android:src="@drawable/tab_home_btn">
        </ImageView>
        <TextView
```

```
        android:id="@+id/textview"
        android:layout_width="wrap_content"
        android:layout_height="wrap_content"
        android:text=" 首页 "
        android:textSize="10sp"
        android:textColor="#ffffff">
    </TextView>
</LinearLayout>
```

第三步：创建 FragmentPage1 文件，并设计界面 1，如图 2.6 所示。具体如代码 CORE0203 所示。

代码 CORE0203：界面 1

```
<!--
View 界面代码
        -->

    <LinearLayout xmlns:android="http://schemas.android.com/apk/res/android"
        xmlns:tools="http://schemas.android.com/tools"
        android:layout_width="match_parent"
        android:layout_height="match_parent"
        android:padding="10dp"
        tools:context=".MainActivity"
        android:orientation="vertical"
        android:background="@drawable/si">
        <TextView
            android:id="@+id/textView1"
            android:layout_width="wrap_content"
            android:layout_height="wrap_content"
            android:layout_marginTop="30dp"
            android:text=" 这是 TextView 显示文本内容 "
            android:textColor="#fff" />
    <Button
            android:id="@+id/button1"
            android:layout_width="wrap_content"
            android:layout_height="wrap_content"
            android:layout_marginTop="24dp"
            android:text=" 按钮 " />
```

```
    <CheckBox
            android:id="@+id/checkBox1"
            android:layout_width="wrap_content"
            android:layout_height="wrap_content"
            android:textColor="#fff"
            android:layout_marginTop="24dp"
            android:text=" 多选方框 " />
    <ToggleButton
            android:id="@+id/toggleButton1"
            android:layout_width="wrap_content"

            android:layout_height="wrap_content"
            android:layout_marginTop="24dp"
            android:text="ToggleButton" />
    <EditText
            android:id="@+id/editText1"
            android:layout_width="wrap_content"
            android:layout_height="wrap_content"
            android:layout_marginTop="24dp"
            android:ems="10" />
    <ImageButton
            android:id="@+id/imageButton1"
            android:layout_width="wrap_content"
            android:layout_height="wrap_content"
            android:layout_marginTop="24dp"
            android:src="@drawable/ic_launcher"
            />
</LinearLayout>
```

第四步：创建 FragmentPage2 文件，并设计界面 2，如图 2.7 所示。具体如代码 CORE0204 所示。

代码 CORE0204：界面 2

```
<!--
Layout 界面代码
      -->
    <?xml version="1.0" encoding="utf-8"?>
    <LinearLayout
```

```
    xmlns:android="http://schemas.android.com/apk/res/android"
android:orientation="vertical"
android:layout_width="match_parent"
android:layout_height="match_parent"
>
<LinearLayout
    android:layout_width="match_parent"
    android:layout_height="0dp"
    android:orientation="horizontal"
    android:layout_weight="1" >
<TextView
android:text="red"
android:gravity="center_horizontal"
android:background="#ff0033"
android:layout_width="wrap_content"
android:layout_height="fill_parent"
android:layout_weight="1"
  />
<TextView
    android:layout_width="wrap_content"
    android:layout_height="fill_parent"
    android:layout_weight="1"
    android:background="#33cc33"
    android:gravity="center_horizontal"
    android:text="green"
 />
<TextView
android:text="blue"
android:gravity="center_horizontal"
android:background="#1281f0"
android:layout_width="wrap_content"
android:layout_height="fill_parent"
android:layout_weight="1"
 />
<TextView
android:text="yellow"
android:gravity="center_horizontal"
android:background="#ff0"
```

```
android:layout_width="wrap_content"
android:layout_height="fill_parent"
android:layout_weight="1"
 />
</LinearLayout>
<LinearLayout
    android:layout_width="match_parent"
    android:layout_height="0dp"
    android:orientation="vertical"
    android:background="#9cf"
    android:layout_weight="1">
<TextView
    android:layout_width="match_parent"
    android:layout_height="wrap_content"
    android:text="Linearout(Vertical)1"
    android:textSize="25dp"
    android:textColor="#fff"
     />
<TextView
    android:layout_width="match_parent"
    android:layout_height="wrap_content"
    android:text="Linearout(Vertical)2"
    android:textSize="25dp"
    android:textColor="#fff"
     />
<TextView
    android:layout_width="match_parent"

    android:layout_height="wrap_content"
    android:text="Linearout(Vertical)3"
    android:textSize="25dp"
    android:textColor="#fff"
     />
<TextView
    android:layout_width="match_parent"
    android:layout_height="wrap_content"
    android:text="Linearout(Vertical)4"
    android:textSize="25dp"
```

```
        android:textColor="#fff"
        />
    </LinearLayout>
    <LinearLayout
        android:layout_width="match_parent"
        android:layout_height="0dp"
        android:orientation="horizontal"
        android:layout_weight="1" >
    <FrameLayout
        android:layout_width="match_parent"
        android:layout_height="match_parent"
        android:layout_weight="1" >
        <TextView
            android:id="@+id/textview1"
            android:layout_width="120dp"
            android:layout_height="120dp"
            android:layout_gravity="center"
            android:background="#ff7f50"
            android:textColor="#fff"/>
        <TextView android:id="@+id/textview2"
            android:layout_width="100dp"
            android:layout_height="100dp"
            android:layout_gravity="center"
            android:background="#ff69b4"
            android:textColor="#fff"/>

        <TextView android:id="@+id/textview3"
            android:layout_width="80dp"
            android:layout_height="80dp"
            android:layout_gravity="center"
            android:background="#ff6347"
            android:textColor="#fff"/>
        <TextView
            android:id="@+id/textview4"
            android:layout_width="60dp"
            android:layout_height="60dp"
            android:layout_gravity="center"
            android:background="#ff4500"
```

```
            android:textColor="#fff" />
        <TextView
            android:id="@+id/textview5"
            android:layout_width="40dp"
            android:layout_height="40dp"
            android:layout_gravity="center"
            android:background="#ff00ff"
            android:textColor="#fff" />
</FrameLayout>
</LinearLayout>
    <!-- 表格 1- 伸展 -->
<LinearLayout
    android:layout_width="match_parent"
    android:layout_height="0dp"
    android:orientation="horizontal"
    android:layout_weight="1" >
    <TableLayout
        android:layout_width="match_parent"
        android:layout_height="match_parent"
        android:background="#99cc33"
        android:padding="4dp"
        android:shrinkColumns="0,1,2" >
        <Button

            android:layout_width="wrap_content"
            android:layout_height="wrap_content"
            android:text=" 我占据一行 "
            />
        <TableRow>
            <Button
                android:layout_width="wrap_content"
                android:layout_height="wrap_content"
                android:text="0000000000000000000000" >
            </Button>
            <Button
                android:layout_width="wrap_content"
                android:layout_height="wrap_content"
                android:text="1111111111111111111111" >
```

```
                </Button>
                <Button
                    android:layout_width="wrap_content"
                    android:layout_height="wrap_content"
                    android:text="22222222222222222222222222" >
                </Button>
            </TableRow>
        </TableLayout>
    </LinearLayout>
</LinearLayout>
```

第五步：创建 FragmentPage3 文件，并设计界面 3，如图 2.8 所示。具体如代码 CORE0205 所示。

代码 CORE0205：界面 3

```
<!--
Dialog 界面代码
    -->
<LinearLayout xmlns:android="http://schemas.android.com/apk/res/android"
    xmlns:tools="http://schemas.android.com/tools"
    android:layout_width="match_parent"
    android:layout_height="match_parent"
    android:padding="10dp"
    tools:context=".MainActivity"
    android:orientation="vertical"
    android:background="@drawable/si">
    <Button
        android:id="@+id/button1"
        android:layout_width="wrap_content"
        android:layout_height="wrap_content"
        android:layout_marginTop="30dp"
        android:text=" 简单提示框 " />
    <Button
        android:id="@+id/button2"
        android:layout_width="wrap_content"
        android:layout_height="wrap_content"
        android:layout_marginTop="24dp"
        android:text=" 带按钮的提示对话框 " />
```

```
        <Button
            android:id="@+id/button3"
            android:layout_width="wrap_content"
            android:layout_height="wrap_content"
            android:layout_marginTop="24dp"
            android:text=" 带多个按钮的提示对话框 " />

        <Button
            android:id="@+id/button4"
            android:layout_width="wrap_content"
            android:layout_height="wrap_content"
            android:layout_marginTop="24dp"
            android:text=" 单选按钮对话框 " />
        <Button
            android:id="@+id/button5"
            android:layout_width="wrap_content"
            android:layout_height="wrap_content"
            android:layout_marginTop="24dp"
            android:text=" 多选按钮对话框 " />
</LinearLayout>
```

第六步:运行程序。运行结果如图 2.6 至图 2.8 所示。

本项目主要讲解了界面布局中一些控件的使用方法以及布局的方式,并且介绍了信息提示框与选择框的知识点以及使用方法。通过本章项目的学习能够熟练地编写布局并掌握提示框用法。

layout	布局
background	背景
color	颜色
drawable	可拉的
create	创建
size	大小

text　　　　文本
gravity　　　重力
center　　　中心

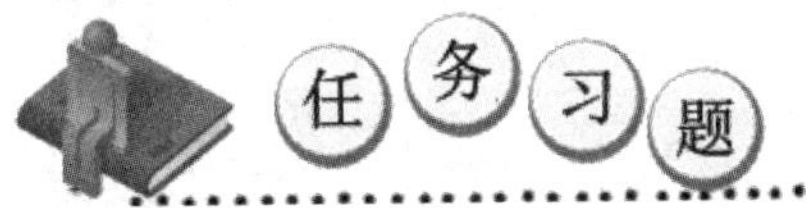

一、选择题

1. 本项目主要介绍了（　　）View 控件。

A.3 个　　B.4 个　　C.5 个　　D.6 个

2. 下列（　　）项不是界面布局五类之一。

A. 相对布局　　B. 绝对布局　　C. 多面布局　　D. 线性布局

3. 线性布局有（　　）种对齐方式。

A.1 种　　B.2 种　　C.3 种　　D.4 种

4. 下列不是 Dialog 的类型的是（　　）。

A.AlertDialog　　B.ProgressDialog

C.DatePickerDialog　　D.setMultiChoiceItems

5. 下列不是安卓布局的是（　　）。

A.FrameLayout　　B.TableLayout　　C.GridView　　D.LinearLayout

二、填空题

1. Dialog 是 Android 开发过程中最常用到的组件之一，它包括以下几种类型：警告对话框、__________、日期选择对话框、__________、时间选择对话框。

2. Android 五大布局有 __________、__________、__________、__________、__________。

3. 如果某列被设为 __________，那么该列的所有单元格宽度可被拉伸。

4. __________ 是按照控件间的相对位置进行布局的。

5. __________ 布局方式与平时使用的 Excel 表格类似。

三、判断题

1. TextView 继承了 View，是 EditView 的父类。（　　）

2. ImageView 是一个文本编译控件。（　　）

3. LinearLayout 里控件的排列方向只能是竖直的。（　　）

4. AlertDialog 的构造方法全部是 Protected，不能直接新建 AlertDialog。（　　）

5. Button 是一种可以设置点击事件的控件。（　　）

四、简答题

1. Dialog 组件的主要作用是什么，它有几种类型？

2. 布局有哪几种，分别有什么特点？

五、上机题

做一个简单的用户登陆界面，要求输入密码时显示“.”，并且在单击登陆时会跳出对话框“用户不存在”，对话框内存在“Yes”和“No”的按钮。

项目三　界面跳转和信息传递

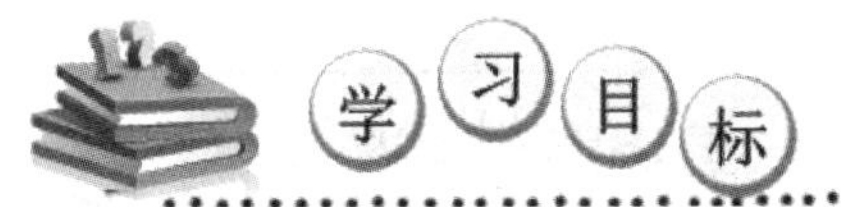

通过实现页面之间的跳转和信息传递，学习 Android 常用界面跳转，Intent 传参的相关知识，了解切换界面以及 Intent 传参的使用方法。在任务实现过程中：

- 掌握页面跳转方法；
- 掌握 Intent 传参的相关知识。

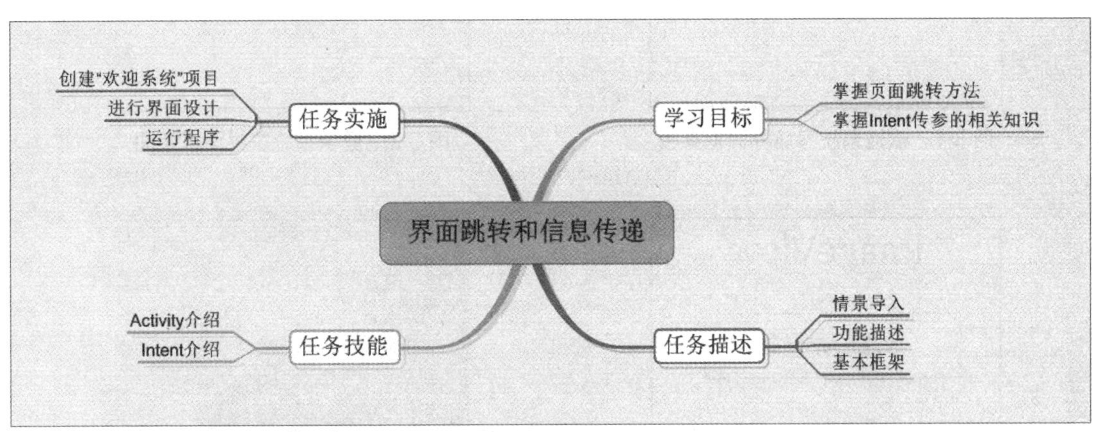

【情景导入】

手机应用程序由多个应用界面组成，当用户实现应用程序的某个功能的时候，往往会需要从当前界面跳转到另一个界面，在 Activity 相互切换时，会有具体的信息传递，为了实现 Activity 跳转以及传参，需要对 Intent 进行进一步了解。本次任务主要实现界面 Intent 跳转和信息传递。

【功能描述】

本任务将设计一款界面跳转的软件。

- 使用线性布局技术来设计登录系统界面；
- 实现欢迎界面跳转到主界面；
- 点击“跳转到音乐室”按钮，跳转到音乐室。

【基本框架】

基本框架如图 3.1 至图 3.3 所示，将框架图转换成的效果如图 3.4 至图 3.6 所示。

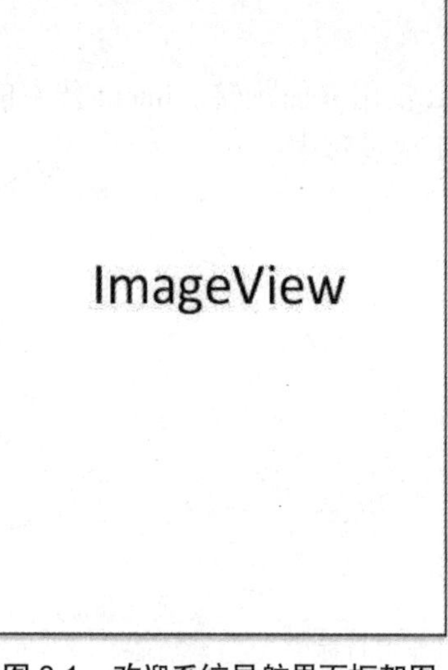

图 3.1 欢迎系统导航界面框架图

图 3.2 登录系统主界面框架图

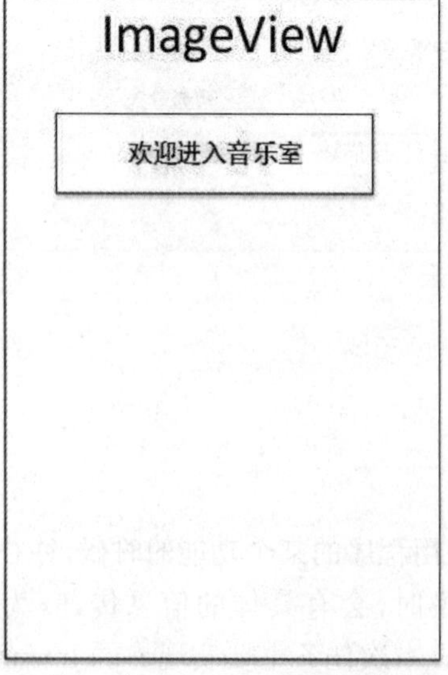

图 3.3 登录系统音乐室界面框架图

图 3.4 欢迎系统导航界面效果图

图 3.5　登录系统主界面效果图

图 3.6　登录系统音乐室界面效果图

技能点 1　Activity 介绍

1　Activity 简介

Acitvity 是一个显示在屏幕上的用户交互界面，是用户可以见到的界面。进一步说，Activity 的每个界面都是独立的，主要是用户体验不同的 Android 应用程序。应用除了可以访问自己的 Activity 外，还可以访问其他 App 的 Activity，这一点会在下面的项目中讲到。

2　Activity 生命周期

Activity 有生命周期，跟人的出生和死亡一样， Activity 实例是由系统创建，并在不同状态期间回调不同的方法。一个最简单的完整的 Activity 生命周期会按照如下顺序回调：

onCreate() → onStart() → onResume() → onPause() → onStop() → onDestroy()。

Android 的生命周期如图 3.7 所示。

onCreate()：生命周期第一个被调用的方法，在创建 Activity 时需重写该方法，在该方法中做一些初始化的操作，例如通过 setContentView() 设置界面布局的资源，初始化组件等。

onStart()：此方法被回调表示 Activity 正在启动，此时 Activity 没在前台显示，无法与用户进行交互。

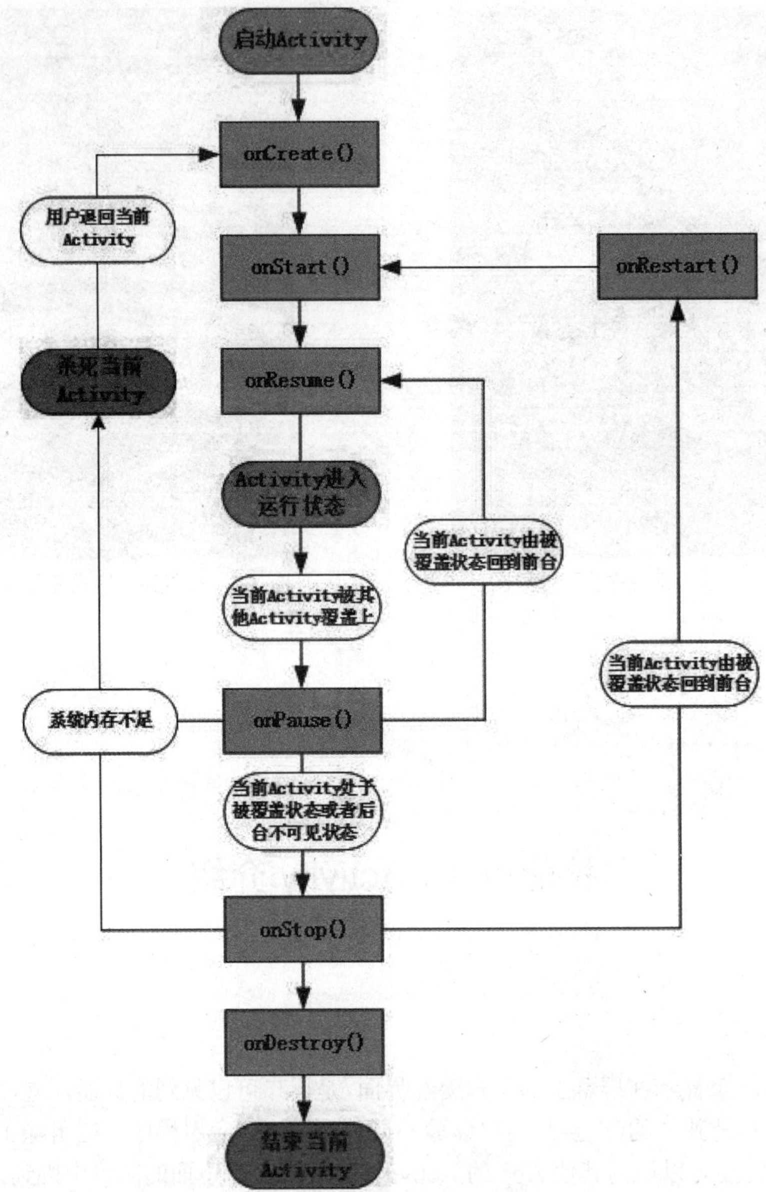

图 3.7　Activity 生命周期

onResume()：此方法被回调表示 Activity 已在前台可见，可与用户交互。从流程图可看出当 Activity 停止后 onPause() 方法和 onStop() 方法被调用，重新回到前台 onResume() 方法也会调用，因此可在 onResume() 方法中初始化信息。

onPause()：表示 Activity 正在停止（Paused 形态），紧接着会回调 onStop() 方法。

onRestart ()：表示 Activity 正在重新启动，当 Activity 由不可见变为可见状态时，该方法被回调。用户打开新 Activity 时，当前 Activity 会被暂停（onPause() 和 onStop() 被执行），然后回到当前 Activity 页面，onRestart() 方法就会被回调。

onDestroy()：此方法被回调时表示 Activity 正在被销毁，该方法是生命周期最后一个执行的方法，一般在此方法中实现回收工作和资源释放。

Activity 失去焦点：如果在 Activity 获得焦点的情况（用户可以见到的时候）下进入其他的 Activity 或应用程序，这时 Activity 会失去焦点。在这个过程中，会依次执行 onPause() 和 onStop() 方法。

Activity 重新获得焦点：如果 Activity 重新获得焦点，可以理解为见到界面，会依次执行 3 个生命周期方法，分别是：onRestart()、onStart() 和 onResume()。

关闭 Activity：当 Activity 关闭系统会运行这 3 个生命周期方法，分别是：onPause()、onStop() 和 onDestroy()。

如果在这 3 个阶段执行生命周期方法的过程中不发生状态的改变则执行 onCreate()、onStart()、onResume()。如果在执行的过程中改变了状态，系统会调用复杂生命周期方法。在执行的过程中可以改变系统的执行过程的生命周期方法有 2 个，分别是 onPause()、onStop()。如果在执行 onPause() 方法的过程中 Activity 重新获得了焦点，然后又失去了焦点，系统将不会再执行 onStop() 方法，而是按着 onPause()、onResume()、onPause() 的顺序执行相应的生命周期方法。

如果程序在执行 onStop() 方法的过程中 Activity 重新获得了焦点（界面再次可见），然后又失去了焦点的话（界面不可见）。将不会执行 onDestroy() 方法，而是顺序依次执行 onStop ()、onRestart ()、onStart ()、onResume ()、onPause ()、onStop()。

Activity 生命周期里可以看出，系统在终止应用程序进程时会依次调用三个方法，即 onPause()、onStop() 和 onDestroy()。onPause() 方法排在了最前面，由此可见 Activity 在失去焦点时就可能被销毁，而 onStop() 和 onDestroy() 方法就可能不会执行。故大多数在 onPause() 方法中保存当前 Activity 状态，这样才能保证在任何时候终止进程时都可以执行保存 Activity 状态的代码。

拓展：想了解或学习更多的 Activity 类以及生命周期使用详情，可扫描下方二维码，获取更多信息。

技能点 2　Intent 介绍

1　Intent 简介

Intent 用于封装程序的“调用意图”。两个 Activity 之间，一般把需要交换的数据封装成 Bundle 对象，然后将 Bundle 对象作为参数传入，就可以实现两个 Activity 之间的数据交换。

Intent 是各种应用程序组件之间“交流”的重要媒介。启动一个 Acitivity、Service 和 BroadcastReceiver（具体见项目八、项目九），Android 均使用统一的 Intent 对象来封装这种将要启动另一个组件的意图。使用 Intent 提供了一致的编程模型。

Intent 可明确指定组件的名称，精确启动某个系统组件，例如精确制定要开启的 Activity 的名字。也可以不指定组件名称，只要能匹配到这个 Intent 的应用都可以接收到，如发送一个拍照 Intent。因为有这种特征的组件很多，所以可以通过在 intent-filter 中配置相应的属性进行处理，这种指定叫模糊指定。

（1）Intent 对象大致包括 7 大属性：ComponentName、Action、Category、Data、Type、Extra、Flag。

- 显式 Intent 是指定了 ComponentName 属性的 Intent 即已经明确了它将要启动哪个组件，反之没有指定 ComponentName 属性的 Intent 被称为隐式 Intent。
- Action 是标识符，当一个 Activity 需要和外部的 Activity 或者广播一起完成某个功能时，就会发出一个 Intent，并在 intent-filter 中添加相应的 Action。在 SDK 中定义了一系列标准动作，如表 3.1 所示。

表 3.1 Action 执行动作

Onstant	Target component	Action
ACTION_CALL	activity	启动一个电话
ACTION_EDIT	activity	显示用户编辑的数据
ACTION_MAIN	activity	作为 Task 中第一个 Activity 启动
ACTION_SYNC	activity	同步手机与数据服务器上的数据
ACTION_BATTERY_LOW	broadcast receiver	电池电量过低警告
ACTION_HEADSET_PLUG	broadcast receiver	拔插耳机警告
ACTION_SCREEN_ON	broadcast receiver	屏幕变亮警告
ACTION_TIMEZONE_CHANGED	broadcast receiver	改变时区警告

- Category 代表 Intent 的种类，Android 上启动 Activity 可以用程序列表、桌面图标、点击 Home 激活桌面等多种方式，Category 则用来标识这些 Activity 的图标会出现在哪些启动的上下文环境里。
- Data 保存需要传递的数据格式，比如：tel://。
- Type 主要是为了对 data 的类型做进一步的说明。一般来说，设置 data 属性为 null，Type 属性才有效，如果 data 属性不设置为 null，系统会自动根据 data 中的协议来分析 data 的数据类型。
- Extra 用来保存过程中传递的数据。
- 通过设置 Flag，可以设置 Activity 是采用哪种启动模式。

（2）Intent 有两种类型状态，分别为显式和隐式。

显式的 Intent：一般这种 Intent 经常用在一个应用中。需要知道要启动的组件名称，如某个 Activity 的包名和类名，在 Intent 中明确地指定了这个组件（Activity）。因为已经明确的知道要启动的组件名称，所以当创建一个显式 Intent 来启动一个 Activity 或者 Service 时，系统会立刻通过你的 Intent 对象启动那个组件。

隐式 Intent：与显示 Intent 最大的区别是隐式 Intent 不知道要启动的组件名称，但是知道 Intent 动作要执行什么动作，比如需要拍照、录像、查看地图等。一般这种 Intent 用在不同的应用之间传递信息。当你创建一个隐式 Intent 时，需要在清单文件中指定 intent-filter，系统会根据 intent-filter 查找匹配的组件。如果你发送的 Intent 匹配到一个 intent-filter，系统会把你的 Intent 传递到对应的组件，并且启动它。如果找到多个匹配的 intent-filter 对应的应用程序，则会弹出一个对话框，该对话框会让你选择由哪个应用程序接收你的 Intent。

2　Intent 使用方法

（1）启动一个 Activity，具体实现方法如下所示：

```
Activity.startActivity(Intent intent); // 启动 Activity
Activity.startActivityForResult(Intent intent, int requestCode);// 带请求 Activity
```

（2）启动 Service，具体实现方法如下所示：

```
Context.startService(Intent service); // 启动 Service
Context.bindService(Intent service, ServiceConnection conn, int flags);// 绑定方法
```

（3）启动 Broadcast（具体讲解见项目九），具体实现方法如下所示：

```
sendBroadcast(Intent intent); // 发普通广播
sendBroadcastAsUser(Intent intent, UserHandle user); // 指定广播发送
sendStickyBroadcast(Intent intent); // 发送滞留广播
sendStickyBroadcastAsUser(Intent intent, UserHandle user);
sendOrderedBroadcast(Intent intent, String receiverPermission); // 发送有序广播
sendOrderedBroadcastAsUser(Intent intent, UserHandle user, String receiverPermis-
sion,BroadcastReceiver resultReceiver,Handler scheduler, int initialCode, String initialData,
Bundle initialExtras);
```

3　Intent 实现步骤

（1）通过 Intent 可以调用并启动其他应用程序，如拨打电话程序。具体实现方法如下所示：

```
public void intentDemo_Call() {
    Intent intent_call = new Intent();                  // 创建一个意图
    intent_call.setAction(Intent.ACTION_CALL);          // 设置意图为打电话
    intent_call.setData(Uri.parse("tel:110"));          // 设置电话号码
    startActivity(intent_call);                         // 启动意图
}
```

（2）使用电话功能需要在 AndroidManifest.xml 文件中添加资源权限方法，具体实现方法如下所示：

```
<uses-permission
        android:name="android.permission.CALL_PHONE"   />
```

（3）Intent 不仅能够调用应用程序，还能实现应用程序内部 Activity 跳转与数据的传递。

● 在 MainActivity 中实现向 SecondaryActivity 发送无请求的意图，具体实现方法如下所示：

```
Intent intent_toSecondary = new Intent();                // 创建一个意图
intent_toSecondary.setClass(this, SecondaryActivity.class);
                                            // 指定跳转到 SecondaryActivity
intent_toSecondary.putExtra("name", "jack");        // 设置传递内容 name
intent_toSecondary.putExtra("age", 23);            // 设置传递内容 age
startActivity(intent_toSecondary);                   // 启动意图
```

● 数据传递后需要在跳转 Activity 中获取数据，具体实现方法如下所示：

```
Intent intent_accept = getIntent();        // 创建一个接收意图
Bundle bundle = intent_accept.getExtras();
                                        // 创建 Bundle 对象，用于接收 Intent 数据
String name = bundle.getString("name");      // 获取 Intent 的内容 name
int age = bundle.getInt("age");                   // 获取 Intent 的内容 age
```

● 在 MainActivity 中实现向 SecondaryActivity 发送带请求码的意图，具体实现方法如下所示：

```
Intent intent_request = new Intent();  // 创建一个意图
intent_request.setClass(this, SecondaryActivity.class);
                                                    // 指定跳转到 SecondaryActivity
startActivityForResult(intent_request, REQUEST_CODE);  // 启动带请求码意图
```

● 接收请求后在意图中填充返回内容并设置返回码，具体实现方法如下所示：

```
Intent intent = getIntent();                         // 创建一个接收意图
intent.putExtra("back", "data of SecondaryActivity"); // 设置意图的内容
setResult(RESULT_CODE, intent);                      // 设置结果码
finish();                  // 结束 SecondaryActivity，并返回 MainActivity
```

（4）当 SecondaryActivity 结束，程序将返回 MainActivity 界面。MainActivity 中的 onActivityResult() 方法将被回调。具体实现方法如下所示：

```
    protected void onActivityResult(int requestCode, int resultCode, Intent data) {
 if(requestCode == REQUEST_CODE && resultCode == SecondaryActivity.RESULT_
CODE) {
            Bundle bundle = data.getExtras();
            String str = bundle.getString("back");
        Toast.makeText(this, " 从 SecondaryActivity 的返回值为：" + str, 0).show();
        }
    }
```

第一步：在 Eclipse 中创建一个 Android 工程，命名为“欢迎系统”，并设计界面。如图 3.4 至图 3.6 所示。

第二步：在 src 文件夹中建立 MainActivity.java 文件和 MenuActivity.java 文件，实现三秒后导航界面跳转到主界面。具体如代码 CORE0301 所示。

代码 CORE0301：跳转

```
// 闪屏跳转界面
public class MainActivity extends Activity {
    private long m_dwSplashTime = 3000;
    private boolean m_bPaused = false;
    private boolean m_bSplashActive = true;
    @Override
    protected void onCreate(Bundle savedInstanceState) {
        super.onCreate(savedInstanceState);
        setContentView(R.layout.activity_main);
        Toast.makeText(getApplicationContext(), " 欢迎进入 ",
                Toast.LENGTH_LONG).show();
        Thread splashTimer = new Thread() {
            public void run() {
                try {
// 延时操作
                    long ms = 0;
                    while (m_bSplashActive && ms < m_dwSplashTime) {
```

```
                        sleep(30);
                        if (!m_bPaused)
                            ms += 30;
                    }
                    startActivity(new android.content.Intent("MainActivity"));
                } catch (Exception ex) {
                    Log.e("Splash", ex.getMessage());
                } finally {
//                  跳转到 MenuActivity 界面
                    Intent intent = new Intent(MainActivity.this,
                            MenuActivity.class);
                    startActivity(intent);
                    finish();
                }
            }
        };
        splashTimer.start();
    }

    protected void onPause() {
        super.onPause();
        m_bPaused = true;
    }
    protected void onResume() {
        super.onResume();
        m_bPaused = false;
    }
}
```

第三步：在 src 文件夹中建立 OtherActivity.java 文件，点击主界面“跳转到音乐室”按钮，实现跳转功能。具体如代码 CORE0302 所示。

代码 CORE0302：跳转到音乐室

```
// 编写单击跳转事件
package com.example.appbasiccoursec_03;
public class MenuActivity extends Activity {
@Override
protected void onCreate(Bundle savedInstanceState) {
```

```
        // TODO Auto-generated method stub
        super.onCreate(savedInstanceState);
        setContentView(R.layout.activity_menu);
        findViewById(R.id.btn_up).setOnClickListener(new OnClickListener() {
            @Override
            public void onClick(View arg0) {
                // TODO Auto-generated method stub
// 从 MenuActivity 跳转到 OtherActivity
                startActivity(new Intent(MenuActivity.this, OtherActivity.class));
            }
        });
}}
```

第四步：运行程序，运行结果如图 3.4 至图 3.6 所示。

【拓展目的】

熟悉并掌握界面跳转及 Intent 传参等技巧。

【拓展内容】

本任务将设计一款利用 Intent 传参跳转的“校园系统”软件。效果如图 3.8、图 3.9 所示。

图 3.8　校园系统主界面

图 3.9　校园系统餐厅界面

【拓展步骤】

（1）设计思路：单击 GridView 条目实现 Intent 跳转与传参。

（2）在 Eclipse 中创建一个 Android 工程，命名为“校园系统”，并设计界面。如图 3.8、图 3.9 所示。

（3）在 src 文件夹中建立 MainActivity.java 文件和 GridViewAdpater.java 文件，并实现 GridView 内容填充。具体如代码 CORE0303 所示。

代码 CORE0303：GridView 内容填充

```
package com.example.appbasiccoursec_03_expand;
public class MainActivity extends Activity {
    private GridView gridview;
    private TextView mTvTemp;
    private TextView mTvHum;
    private GridViewAdpater gridViewAdpater;
    @Override
    protected void onCreate(Bundle savedInstanceState) {
        super.onCreate(savedInstanceState);
        setContentView(R.layout.activity_main);
        initView();
    }
/**
 * 1 此处填写 Gridview 内容填充
 */
    private void initView() {
        gridview = (GridView) findViewById(R.id.mygridView);
// 给 Gridview 设置适配器
        gridViewAdpater = new GridViewAdpater(this);
        gridview.setAdapter(gridViewAdpater);
    }
    @Override
    protected void onDestroy() {
        super.onDestroy();
    }
}
/**
 * 2 此处填写适配器实现代码
 */
public class GridViewAdpater extends BaseAdapter {
```

```
    private int[] color = new int[] { R.color.main_blue, R.color.main_blue0,
            R.color.main_blue1, R.color.main_lgreen, R.color.main_orange,
            R.color.main_orangered };
    private String[] str = new String[] { " 音乐室 ", " 游泳馆 ", " 篮球馆 ",
            " 实训室 ", " 宿舍 ", " 餐厅 " };
    private Context mContext;
    public GridViewAdpater(Context mContext) {
        // TODO Auto-generated constructor stub
// 获取上下文
        this.mContext = mContext;
    }
    @Override
    public int getCount() {
        // TODO Auto-generated method stub
// 返回字符串长度
        return str.length;
    }
    @Override
    public String getItem(int arg0) {
        // TODO Auto-generated method stub
        return str[arg0];
    }
    @Override
    public long getItemId(int arg0) {
        // TODO Auto-generated method stub
        return arg0;
    }
    @Override
    public View getView(int position, View convertView, ViewGroup parent) {
        // TODO Auto-generated method stub
        ViewHolder vh = null;
        if (convertView == null) {
            vh = new ViewHolder();
// 获取视图界面并且实现初始化界面
            convertView = LayoutInflater.from(mContext).inflate(
                R.layout.gridview, null);
            vh.lin = (LinearLayout) convertView.findViewById(R.id.lin);
            vh.text = (TextView) convertView.findViewById(R.id.text);
```

```
                convertView.setTag(vh);
            } else {

                vh = (ViewHolder) convertView.getTag();
            }
    // 根据索引设置背景图片以及标题
            vh.lin.setBackgroundResource(color[position]);
            vh.text.setText(str[position]);
            return convertView;
        }
    // 创建命名
        class ViewHolder {
            TextView text;
            LinearLayout lin;
        }
    }
```

（4）在 src 文件夹中建立 OtherExhibition.java 文件，实现 GirdView 单击条目事件跳转到 OtherExhibition 界面并实现 Intent 传参。具体如代码 CORE0304 所示。

代码 CORE0304：Intent 传参

```
/**
    * 3GridView 点击跳转跳转传参事件
    */
    gridview.setOnItemClickListener(new OnItemClickListener() {
                @Override
                public void onItemClick(AdapterView<?> arg0, View arg1, int arg2,
                        long arg3) {
                    // TODO Auto-generated method stub
                    Intent intent = new Intent();
                    switch (arg2) {
                    case 0:
                    intent.setClass(MainActivity.this, OtherExhibition.class);
                        break;
                    case 1:
                        intent.setClass(MainActivity.this, OtherExhibition.class);
                        break;
                    case 2:
```

```
                        intent.setClass(MainActivity.this, OtherExhibition.class);
                        break;
                    case 3:
                        intent.setClass(MainActivity.this, OtherExhibition.class);
                        break;
                    case 4:
                        intent.setClass(MainActivity.this, OtherExhibition.class);
                        break;
                    case 5:
                        intent.setClass(MainActivity.this, OtherExhibition.class);
                        break;
                    }
//                  设置传递参数内容并且进行跳转
                    intent.putExtra("type", arg2);//key="type",value=arg2
                    intent.putExtra("name", gridViewAdpater.getItem(arg2));
                    startActivity(intent);
                }
            });
```

（5）根据传递的信息更改 OtherExhibition 界面，具体如代码 CORE0305 所示。

代码 CORE0305：接收信息更新界面

```
public class OtherExhibition extends Activity {
    @SuppressLint("ResourceAsColor")

    @Override
    protected void onCreate(Bundle savedInstanceState) {
        // TODO Auto-generated method stub
        super.onCreate(savedInstanceState);
        setContentView(R.layout.other);
/**
 * 4 此处填写获取传递参数值并修改显示界面
 */
 // 获取 Intent 传递参数
        String name = getIntent().getStringExtra("name");
        int type = getIntent().getIntExtra("type", -1);
        ((TextView) findViewById(R.id.name)).setText(name);
        Toast.makeText(this, " 欢迎进入 "+name, 0).show();
```

```
// 初始化界面
            LinearLayout linear=(LinearLayout) findViewById(R.id.img);
            RelativeLayout title=(RelativeLayout) findViewById(R.id.title);
            switch (type) {
            case 0:
// 设置背景及标题颜色
                linear.setBackgroundResource(R.drawable.music);
                title.setBackgroundColor(getResources().getColor(R.color.main_blue));
                break;
            case 1:
                linear.setBackgroundResource(R.drawable.bg_swim);
                title.setBackgroundColor(getResources().getColor(R.color.main_blue0));
                break;
            case 2:
                linear.setBackgroundResource(R.drawable.bg_basketball);
                title.setBackgroundColor(getResources().getColor(R.color.main_blue1));
                break;
            case 3:
                linear.setBackgroundResource(R.drawable.classroom);

                title.setBackgroundColor(getResources().getColor(R.color.main_lgre
en));
                break;
            case 4:
                linear.setBackgroundResource(R.drawable.room);
                title.setBackgroundColor(getResources().getColor(R.color.main_oran
ge));
                break;
            case 5:
                linear.setBackgroundResource(R.drawable.bg_mess);
                title.setBackgroundColor(getResources().getColor(R.color.main_orang-
ered));
                break;
            default:
                break;
            }
        }
    }
```

本项目主要介绍了用 Android 实现 Activity 之间的跳转和页面信息传递的相关知识和技能，通过本项目的学习，使学生对页面跳转和传参有了更深的了解，并且能够熟练地使用 Intent 跳转传参。

intent	意图
head	头文件
text	文本
color	颜色
background	背景
background-color	背景颜色
select	选择
linearLayout	线性布局
relative	相对
horizontal	水平的

一、选择题

1. Intent 的作用的是(　　)。

A. 是一段长的生命周期，没有用户界面的程序，可以保持应用在后台运行，而不会因为切换页面而消失 service

B. Intent 是连接四大组件的纽带，可以实现界面间切换，可以包含动作和动作数据

C. 实现应用程序间的数据共享 contentprovider

D. 处理一个应用程序整体性的工作

2. Activity 在运行的时候大致会经过四个状态，下列那一个不是(　　)。

A. 活动状态　　B. 暂停状态　　C. 销毁状态　　D. 跳转状态

3. Intent 对象大致包括 7 大属性，下列哪一项不在这些属性之内(　　)。

A.Action　　B.Category　　C.Data　　D.onResume

4. Intent 有两种类型状态，下面哪一项是符合这两种状态的(　　)。

A. 显示状态　　B. 多线状态　　C. 回转状态　　D. 联结状态

5. Intent 传递数据时，下列的那个选项不可作为数据被传递(　　)。

A.Serializable B.charsequence C.Parcelable D.xml

二、填空题

1. Intent 开启一个 Activity 用的方法是 ________。
2. Intent 的意图类型：________。
3. Intent 对象的属性中，________ 保存需要传递的额外数据。
4. Intent 设置 Activity 启动模式的方式是 ________。
5. Intent 常见的应用请写三种：________、________、________。

三、判断题

1. startActivity 是 Intent 的方法。 ( )
2. Intent 分为显示意图和隐式意图。 ( )
3. Intent 不可以传递对象。 ( )
4. StartActivity(MainActivity,TextActivity)。 ( )
5. Intent 是 Android 的四大组件。 ( )

四、简答题

1. Intent 可以传递的参数有哪些？
2. Intent 的作用？

五、上机题

设计一个短信发送界面，并实现短信发送功能。

项目四　规范应用资源

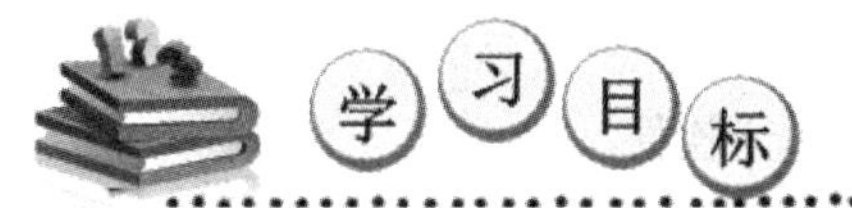

通过实现“天津美景系统”应用资源的使用，学习如何合理使用应用资源的相关知识编写应用资源，了解应用资源的使用方法。在任务实现过程中：

- 了解 Android 应用资源的类型；
- 掌握 Android 中数组的使用方法；
- 了解界面设计的样式；
- 掌握国际化的相关知识理念和技能。

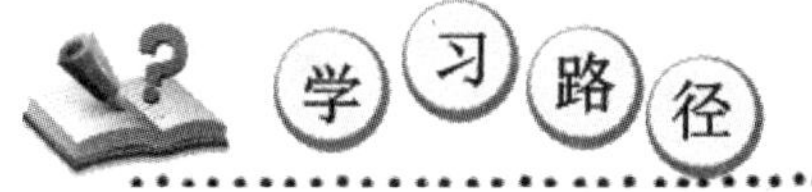

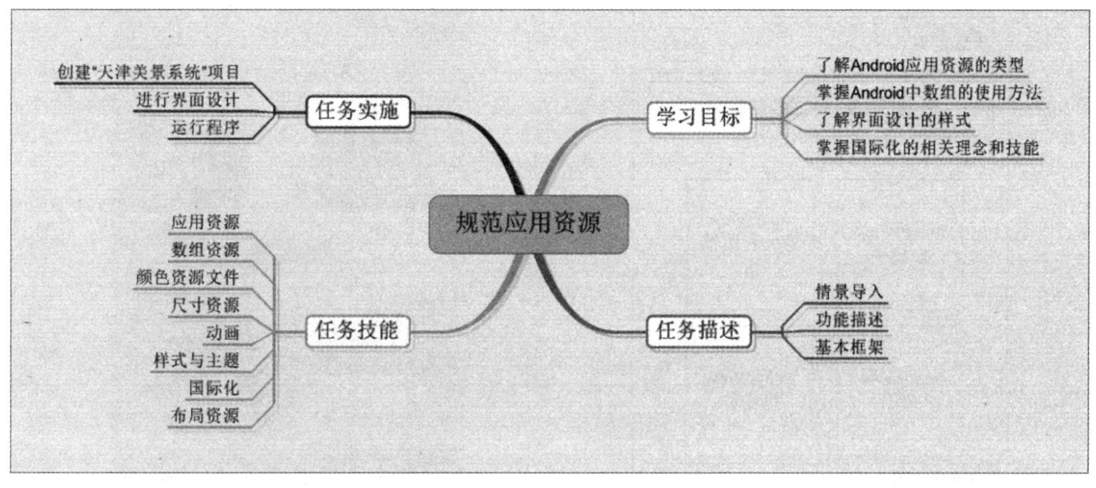

【情景导入】

各式各样的 App 已经进入了人们的生活，用户对于 App 的实用性以及美观程度也越来越重视。应用规范资源可以便捷地对界面进行设计、美化。本项目使用应用资源主要实现对 Android 程序界面进行美化以及填充界面信息。

【功能描述】

本任务将设计一款使用应用资源开发的“天津美景系统”软件。

- 在 colors.xml 中编写界面所需颜色；
- 在 mystyle.xml 中编写界面所需的样式主题；
- 在 arrays.xml 编写数组信息；
- 在 strings.xml 中编写文字内容；
- 使用 LinearLayout 设计登录系统界面；
- 在 ListView 中填充数据；
- 在 menu 文件下 main.xml 中编写系统按键信息；
- 在点击系统按键实现语言种类切换。

【基本框架】

基本框架如图 4.1 所示，将框架图转换成的效果如图 4.2 所示。

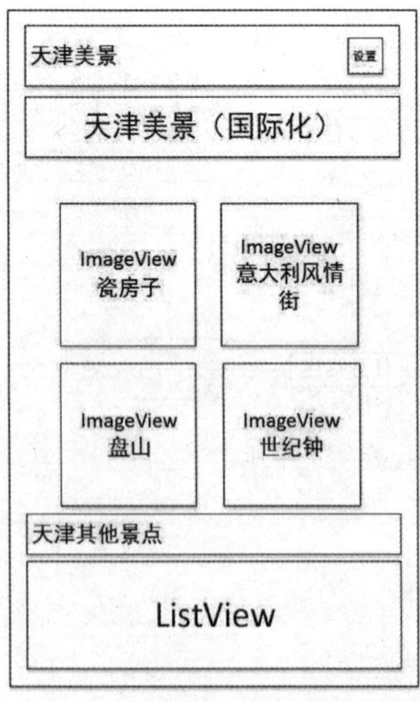

图 4.1 天津美景系统主界面框架图

图 4.2 天津美景系统主界面效果图

技能点 1　应用资源

应用资源指与 UI 相关的资源，如 UI 布局、字符串和图片等。代码和资源分开使应用程序只需编译一次，且能够支持不同的 UI 的布局。这种特性使应用程序运行时可以适应不同的屏幕大小，还可以适应不同国家的语言等。

Android 应用资源分为两大类，分别是 assets 和 res。其中 assets 类资源存放在 assets 的子目录下。在 assets 中保存的资源文件最终会被打包在 apk 中，需根据指定文件名进行使用。res 类资源存储在工程根目录 res 子目录下。在 res 中的资源会被赋予资源 ID，在程序中通过 ID 对资源进行访问。res 类资源可根据不同用途划分成以下九类（表 4.1）。

表 4.1　资源文件存储方式

资源类型	所需的目录	文件名规范	适用的关键 XML 元素
字符串	/res/values/	strings.xml	<string>
字符串数组	/res/values/	arrays.xml	<string-array>
颜色值	/res/values/	colors.xml	<color>
尺寸	/res/values/	dimens.xml	<dimen>
简单 Drawable 图形	/res/values/	drawables.xml	<drawable>
动画序列（补间）	/res/anim/	fancy_anim.xml	<set>、<alpha>、<scale>、<rotate>
样式和主题	/res/values/	themes.xml	<style>
菜单文件	/res/menu/	my_menu.xml	<menu>
XML 文件	/res/xml/	some.xml	由开发人员定义

拓展：想了解或学习图片（Drawable）资源、属性（Attribute）资源，可扫描下方二维码，获取更多信息。

技能点 2　数组资源

1　数组资源介绍

Android 中数组（array）的定义方式有两种，第一种可以直接在 Android 代码中声明，第二种可以在 res/values 目录下新建一个 xml 文件，对数组资源进行声明。字符数组有两种声明方式，分别是 String[] 和 List<String>。

在实际开发中，最好将数据存放在资源文件中，这样来实现程序的逻辑代码与数据分离，便于项目的管理，减少对 Java 代码的修改。

2　数组资源使用方法

Android 规定存放数组的文件必须在 res/values 文件夹下创建，推荐该文件名 arrays.xml。以下代码定义了含有四个直辖市名称的字符串数组，数组名是 citys，数组元素在 <item> 标签中存放。

在 Android 中提供了 Resource 类，可以调用数组中的内容，通过该类提供的方法可以方便地获取资源中的数据，如资源中定义的数组。具体实现方法如下所示：

```
<!-- 字符串数组存储在 /res/values/arrays.xml 文件中，格式如下所示：-->
<?xml version="1.0" encoding="utf-8"?>
<resources>
    <string-array name="citys ">
        <item> 北京 </item>
        <item> 上海 </item>
        <item> 天津 </item>
        <item> 重庆 </item>
    </string-array>
</resources>
<!-- 获取字符数组内容需要通过如下方式 -->String strs[] = getResources().getSt-
ringArray(R.array. citys);
```

技能点 3　颜色资源文件

1　颜色资源介绍

在使用 UI 界面，并对其进行布局的时候，使用的控件可以通过“android:textColor”和“android:background”这两种属性，分别给文字和背景附上颜色。

（1）颜色定义方式

颜色表示：颜色通过 红（red）绿（green）蓝（blue）三种颜色，以及透明度（alpha）来表示的。

颜色开头：颜色值总是以 # 开头，无透明度，如果没有 alpha 值，默认完全不透明。

（2）颜色定义形式

- #RGB：红 绿 蓝 三原色值，每个值分 16 个等级，最小为 0，最大为 F；
- #ARGB：透明度 红 绿 蓝 值，每个值分 16 个等级，最小为 0，最大为 F；
- #RRGGBB：红 绿 蓝 三原色值，每个值分 256 个等级，最小为 0，最大为 FF；
- #AARRGGBB：透明度 红 绿 蓝 值，每个值分 256 个等级，最小为 0，最大为 FF。

十六进制颜色值如图 4.3 所示。

2　颜色资源使用

数组元素在 <color> 标签中存放，具体实现方法如下所示：

```
<!--color 文件 -->
<?xml version="1.0" encoding="utf-8"?>
<resources>
<color name="black">#120A2A</color>
<color name="red">#FF4000</color>
 <color name="yellow">#FFFF00</color>
 <color name="burlywood">#1281f0</color>
</resources>
<!--layout 文件下 -->
<TextView
    android:id="@+id/textView3"
    android:layout_width="wrap_content"
    android:layout_height="wrap_content"
    android:text="@string/yi"
    android:textColor="@color/red" />
```

aliceblue #F0F8FF	antiquewhite #FAEBD7	aqua #00FFFF
aquamarine #7FFFD4	azure #F0FFFF	beige #F5F5DC
bisque #FFE4C4	black #000000	blue #0000FF
blueviolet #8A2BE2	brown #A52A2A	burlywood #DEB887
cadetblue #5F9EA0	chartreuse #7FFF00	chocolate #D2691E
coral #FF7F50	cornflowerblue #6495ED	cornsilk #FFF8DC
crimson #DC143C	cyan #00FFFF	darkblue #00008B
darkcyan #008B8B	darkgoldenrod #B8860B	darkgray #A9A9A9
darkgreen #006400	darkkhaki #BDB76B	darkmagenta #8B008B
darkolivegreen #556B2F	darkorange #FF8C00	darkorchid #9932CC
darkred #8B0000	darksalmon #E9967A	darkseagreen #8FBC8F
darkslateblue #483D8B	darkslategray #2F4F4F	darkturquoise #00CED1
deepskyblue #00BFFF	dimgray #696969	dodgerblue #1E90FF
firebrick #B22222	floralwhite #FFFAF0	forestgreen #228B22
fuchsia #FF00FF	gainsboro #DCDCDC	ghostwhite #F8F8FF
gold #FFD700	goldenrod #DAA520	gray #7F7F7F
green #008000	greenyellow #ADFF2F	honeydew #F0FFF0
hotpink #FF69B4	indianred #CD5C5C	indigo #4B0082
ivory #FFFFF0	khaki #F0E68C	lavender #E6E6FA
lavenderblush #FFF0F5	lawngreen #7CFC00	lemonchiffon #FFFACD
lightblue #ADD8E6	lightcoral #F08080	lightcyan #E0FFFF
lightgreen #90EE90	lightgrey #D3D3D3	lightpink #FFB6C1
lightsalmon #FFA07A	lightseagreen #20B2AA	lightskyblue #87CEFA
lightslategray #778899	lightsteelblue #B0C4DE	lightyellow #FFFFE0
lime #00FF00	limegreen #32CD32	linen #FAF0E6

图 4.3 颜色列表

技能点 4　尺寸资源

1　尺寸资源介绍

尺寸实质就是控件的长宽高，以及页边距的值。如何将控件合理地布置在 UI 界面中，是十分重要的。调整尺寸经常用到的一些属性如表 4.2 所示。

表 4.2　尺寸属性

常用属性	描述
android:layout_width	调整宽度
android:layout_height	调整高度
android:layout_marginLeft	调整左页边距
android:layout_marginTop	调整上页边距
android:layout_marginRight	调整右页边距

2　尺寸使用方法

尺寸存储格式及获取尺寸内容方式，具体实现方法如下所示：

```
<!-- 尺寸存储在 /res/values/dimens.xml 文件中 --><?xml version="1.0" encoding="utf-8"?>
<resources>
  <dimen name="txt_app_title">22sp</dimen>
  <dimen name="font_size_10">10sp</dimen>
  <dimen name="font_size_12">12sp</dimen>
  <dimen name="font_size_14">14sp</dimen>
  <dimen name="font_size_16">16sp</dimen>
</resources>
<!-- 取尺寸使用下列代码：-->
float myDimen = getResources().getDimension(R.dimen.dimen 标签 name 属性的名字);
```

技能点 5 动画

1 动画介绍

Android 中将动画分为帧动画、补间动画、属性动画三种（具体见项目七），这些动画都可以使用 XML 文件定义。

2 动画实现方法

动画存储在 /res/anim/ 目录中，也可以将其写在 Android 代码中，XML 动画具体实现方法如下所示：

```
<?xml version="1.0" encoding="UTF-8"?>
<set xmlns:android="http://schemas.android.com/apk/res/android">
   <rotate
      android:interpolator="@android:anim/accelerate_decelerate_interpolator"
      android:fromDegrees="300"
      android:toDegrees="-360"
      android:pivotX="10%"
      android:pivotY="100%"
      android:duration="10000" />
</set>
<!-- 旋转控制动画效果 rotate
         fromDegrees 为旋转起始角度
         toDegrees 为旋转结束角度
         （pivotX，pivotY）起始点坐标
         Duration 为动画时间
-->
<!-- 加载动画资源 -->
         Animation animation = AnimationUtils.loadAnimation(this, R.anim.transla-
tion_anim);
<!-- 开启动画 -->
         imageView.startAnimation(animation);
```

Android 代码中具体实现方法如下所示：

```
/** 设置旋转动画 */
final
    RotateAnimation animation =new RotateAnimation(0f,360f,Animation.RELA-
TIVE_TO_SELF,
0.5f,Animation.RELATIVE_TO_SELF,0.5f);
animation.setDuration(3000);// 设置动画持续时间
/** 常用方法 */
//animation.setRepeatCount(int repeatCount);// 设置重复次数
//animation.setFillAfter(boolean);// 动画执行完后是否停留在执行完的状态
//animation.setStartOffset(long startOffset);// 执行前的等待时间
start.setOnClickListener(new OnClickListener() {
public void onClick(View arg0) {
image.setAnimation(animation);
/** 开始动画 */
animation.startNow();
```

技能点 6　样式与主题

1　样式与主题介绍

样式是用来指定视图或窗口的外观和格局的一组属性集合。主题是用来指定整个布局样式的属性集合。例如定义属性 layout_width、layout_height 等样式可以用来指定宽度、高度，还可以指定字体大小、背景颜色等。样式以独立的资源文件形式存放在 XML 文件中。主题与样式唯一不同的地方是样式只是应用于某个控件，而主题是全屏应用于“Activity”。在使用手机或者应用的过程中，经常会需要应用全屏的效果，也就是隐藏上面的状态栏和标题栏。例如在一些浏览器或者手机游戏 App 中都有全屏的效果，而全屏的功能再加上其中被样式美化的控件所形成的效果就是主题。

样式定义在一个单独的 XML 资源文件中。该 XML 文件位于 res/values/ 目录中，必须使用样式文件中 <resources> 作为根节点并使用 .xml 扩展名，XML 文件名称是任意的。

2　样式与主题使用方法

（1）可以定义每个文件中使用的多种样式 <style> 标签，但要使用唯一名称标识该样式。Android 样式属性设置使用的 <item> 标签，具体实现方法如下所示：

```
<?xml version="1.0" encoding="utf-8"?>
            <resources>
              <style name="CustomFontStyle">
                <item name="android:layout_width">fill_parent</item>
                <item name="android:layout_height">wrap_content</item>
                <item name="android:capitalize">characters</item>
                <item name="android:typeface">monospace</item>
                <item name="android:textSize">12pt</item>
                <item name="android:textColor">#00FF00</item>/>
              </style>
            </resources>
```

（2）<item> 中的值可以是多种类型值，可以是一个关键字串，也可以是十六进制的颜色。使用样式定义后，可用在 XML 布局文件中使用样式属性，具体实现方法如下所示：

```
<?xml version="1.0" encoding="utf-8"?>
                  <LinearLayout    xmlns:android="http://schemas.android.com/apk/res/
android"
                  android:layout_width="fill_parent"
                  android:layout_height="fill_parent"
                  android:orientation="vertical" >
                  <TextView
                  android:id="@+id/text_id"
                  style="@style/CustomFontStyle"
                   android:text="@string/hello_world" />
                  </LinearLayout>
```

技能点 7　国际化

1　国际化介绍

Internationalization（国际化）的简称 是 i18n，因为在 i 和 n 之间有 18 个字符，localization（本地化），简称 L10n，用“语言 _ 地区”的形式说明一个地区的语言时，如 zh_CN，zh_TW。

2　国际化实现

Android 对 i18n 和 L10n 提供了非常好的支持。Android 是通过对不同 resource 的命名来达到国际化，而没有 API 来提供国际化。其实现方法是建立 values-zh,values-en 文件夹，在这个文件内放置 strings.xml，根据不同的语言编写不同的信息，代码如下所示：

```
<string name="english">English</string>
<string name="simple_chinese"> 简体中文 </string>
<string name="traditional_chinese"> 繁體中文 </string>
```

技能点 8　布局资源

1　布局资源介绍

布局资源（layout）是 Android 中最常用的应用资源，本项目详细介绍布局资源使用方法。如果一个界面想在手机或者虚拟机中展现出来，必须有界面的布局文件，以 XML 文件的形式存在。

这里先介绍一下 Android 的用户界面，布局问题也会导致程序运行失败。在一个 Android 应用程序中，用户界面要通过 Views 和 ViewGroup 对象构建。Android 中有很多种 Views 和 ViewGroup，它们都继承自 View 类。Views 对象是 Android 平台上表示用户界面的基本单元，继承关系图如图 4.4 所示。

要将视图层次树呈现到屏幕上，就必须调用 setContentView() 方法并且传递根节点对象的引用。

2　布局资源使用方法

下面先实现怎么在 XML 文件里写布局。下面的 XML 布局文件使用了纵向的 LinearLayout，此布局中有一个 TextView 和一个 Button 控件。

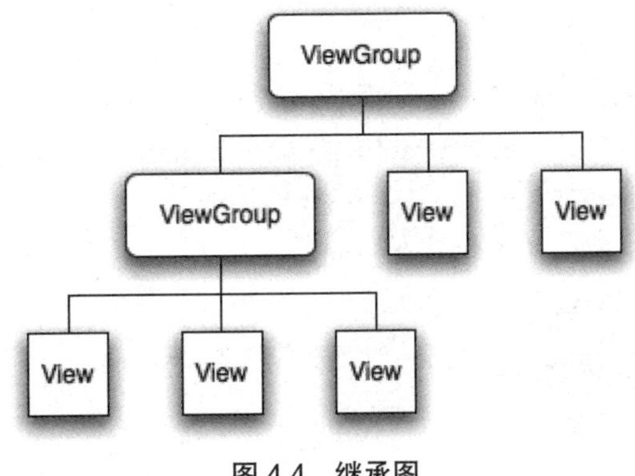

图 4.4　继承图

```
<?xml version="1.0" encoding="utf-8"?>
<LinearLayout xmlns:android="http://schemas.android.com/apk/res/android"
        android:layout_width="fill_parent"
        android:layout_height="fill_parent"
        android:orientation="vertical" >
    <TextView
        android:id="@+id/text"
        android:layout_width="wrap_content"
        android:layout_height="wrap_content"
        android:text="Hello, I am a TextView" />
    <Button
        android:id="@+id/button"
        android:layout_width="wrap_content"
        android:layout_height="wrap_content"
        android:text="Hello, I am a Button" />
</LinearLayout>
```

任务实施

第一步：在 Eclipse 中创建一个 Android 工程，命名为“天津美景系统”。具体如代码 CORE0401 所示。

代码 CORE0401：strings.xml

```
<!--
此处填写风景区名称 -->
<?xml version="1.0" encoding="utf-8"?>
<resources>
    <string name="app_name"> 天津美景 </string>
    <string name="hello_world">Tianjin beauty</string>
    <string name="action_settings">Settings</string>
    <string name="english">English</string>
    <string name="simple_chinese"> 简体中文 </string>
    <string name="traditional_chinese"> 繁體中文 </string>
</resources>
```

第二步：在 colors.xml 中添加界面所需颜色，具体如代码 CORE0402 所示。

代码 CORE0402：colors.xml

```
<!--
  此处填写颜色代码 -->
<?xml version="1.0" encoding="utf-8"?>
<resources>
<color name="black">#120A2A</color>
<color name="red">#FF4000</color>
<color name="yellow">#FFFF00</color>
<color name="burlywood">#1281f0</color>
  </resources>
```

第三步：在 mystyle.xml 中添加界面所需样式主题，具体如代码 CORE0403 所示。

代码 CORE0403：mystyle.xml

```
<!--
此处填写样式代码 -->

<resources>
  <style name="TextStyle">
     <item name="android:textSize">30sp</item>
    <item name="android:textColor">#FF4000</item>
  </style>
<style name="MyTheme">
<item name="android:background">#D9D9D9</item>
</style>
 <!-- Application theme. -->
 <style name="AppBaseTheme" parent="android:Theme.Holo.Light">
    <!-- API 11 theme customizations can go here. -->
  </style>
</resources>
```

第四步：在 arrays.xml 中添加数组内容，具体如代码 CORE0404 所示。

代码 CORE0404：arrays.xml

```
<?xml version="1.0" encoding="utf-8"?>
<resources>
  <string name="ci"> 瓷房子 </string>
  <string name="yi"> 意大利风情街 </string>
  <string name="ji"> 盘山 </string>
```

```
    <string name="shi"> 世纪钟 </string>
    <string-array name="citys">
        <item> 水上公园 </item>
        <item> 天津眼 </item>
        <item> 天塔 </item>
        <item> 外滩 </item>
    </string-array>
</resources>
```

第五步：创建 values-zh-rCN 文件与 values-zh-rTW 文件并且在文件中分别创建。在 strings.xml 中编写所需内容，具体如代码 CORE0405 和 CORE0406 所示。

代码 CORE0405：values-zh-rCN

```
<?xml version="1.0" encoding="utf-8"?>
<resources>
    <string name="hello_world"> 天津美景！ </string>
</resources>
```

代码 CORE0406：values-zh-rTW

```
<?xml version="1.0" encoding="utf-8"?>
<resources>
    <string name="hello_world"> 天津美景，繁體字 ！ </string>
</resources>
```

第六步：合理利用应用资源设计界面。如图 4.2 所示，具体如代码 CORE0407 所示。

代码 CORE0407：主界面

```
<LinearLayout xmlns:android="http://schemas.android.com/apk/res/android"
    xmlns:tools="http://schemas.android.com/tools"
    android:layout_width="match_parent"
    android:layout_height="match_parent"
    android:background="@drawable/rootblock_default_bg"
    android:orientation="vertical"
    tools:context=".MainActivity" >
<LinearLayout
    android:layout_width="match_parent"
    android:layout_height="wrap_content"
    android:gravity="center_horizontal"
```

```
    android:orientation="vertical"
    android:layout_weight="1" >
    <TextView
      android:id="@+id/textView1"
      android:layout_width="wrap_content"
      android:layout_height="wrap_content"
      style="@style/TextStyle"
      android:text="@string/hello_world"
      />
  </LinearLayout>
<LinearLayout
    android:layout_width="match_parent"
    android:layout_height="162dp"
    android:layout_weight="4"
    android:orientation="horizontal" >
    <LinearLayout
      android:layout_width="match_parent"
      android:layout_height="match_parent"
      android:orientation="vertical"
      android:layout_weight="1"
      android:gravity="center"
    >
    <ImageView
      android:id="@+id/imageView1"
      android:layout_width="120dp"
      android:layout_height="100dp"
      android:src="@drawable/cifangzi" />
       <TextView
      android:id="@+id/textView2"
      android:layout_width="wrap_content"
      android:layout_height="wrap_content"
      android:text="@string/ci"/>
</LinearLayout>
 <LinearLayout
      android:layout_width="match_parent"
      android:layout_height="match_parent"
      android:orientation="vertical"
      android:layout_weight="1"
```

```
        android:gravity="center"
      >
      <ImageView
        android:id="@+id/imageView2"
        android:layout_width="110dp"
        android:layout_height="92dp"
        android:scaleType="fitXY"
        android:src="@drawable/fengqingjie" />
          <TextView
        android:id="@+id/textView3"
        android:layout_width="wrap_content"
       android:layout_height="wrap_content"
       android:text="@string/yi"
       android:textColor="@color/red" />
    </LinearLayout>
      </LinearLayout>
    <LinearLayout
        android:layout_width="match_parent"
        android:layout_height="162dp"
        android:layout_weight="4"
        android:orientation="horizontal"
          >
      <LinearLayout
        android:layout_width="match_parent"
        android:layout_height="match_parent"
        android:orientation="vertical"
        android:layout_weight="1"
        android:gravity="center"
      >
       <ImageView
        android:id="@+id/imageView3"
        android:layout_width="120dp"
        android:layout_height="100dp"
        android:src="@drawable/panshan" />
        <TextView
        android:id="@+id/textView4"
        android:layout_width="wrap_content"
        android:layout_height="wrap_content"
```

```
        android:layout_marginBottom="16dp"
        android:text="@string/ji"
        android:textColor="@color/yellow"
        />
</LinearLayout>
 <LinearLayout
        android:layout_width="match_parent"
        android:layout_height="match_parent"
        android:orientation="vertical"
        android:layout_weight="1"
        android:gravity="center"
    >
  <ImageView
        android:id="@+id/imageView2"
        android:layout_width="120dp"
        android:layout_height="100dp"
        android:scaleType="fitXY"
        android:src="@drawable/shijizhong" />
         <TextView
        android:id="@+id/textView5"
        android:layout_width="wrap_content"
        android:layout_height="wrap_content"
        android:text="@string/shi"
        android:textColor="@color/black"
        />
</LinearLayout>
</LinearLayout>
 <LinearLayout
        android:layout_width="match_parent"
        android:layout_height="wrap_content"
        android:orientation="vertical"
        android:layout_weight="1"
        android:gravity="center"
    >
<TextView
       android:layout_width="match_parent"
        android:layout_height="wrap_content"
        android:text=" 天津其他景点 "
```

```
        android:textColor="@color/black"
        android:textSize="25dp"
    />
</LinearLayout>
<View
        android:layout_width="match_parent"
        android:layout_height="0.5dp"
        android:background="@color/black"
    />
    <LinearLayout
      android:layout_width="match_parent"
        android:layout_height="wrap_content"
        android:orientation="vertical"
        android:layout_weight="1"
        android:gravity="center"
    >
    <ListView
        android:id="@+id/listview"
        android:layout_width="match_parent"
        android:layout_height="150dp"
         >
</ListView>
</LinearLayout>
</LinearLayout>
```

第七步：将 arrays.xml 中信息填充到 ListView 中。具体如代码 CORE0408 所示。

代码 CORE0408：填充 ListView

```
listView = (ListView)findViewById(R.id.listview);
        data = getResources().getStringArray(R.array.citys);
        arrayAdapter = new ArrayAdapter<String>(MainActivity.this,android.R.layout.
simple_list_item_1, data);
        listView.setAdapter(arrayAdapter);
```

第八步：在 menu 下 main.xml 中编辑系统按键信息。具体如代码 CORE0409 所示。

代码 CORE0409：系统按键信息

```
<menu xmlns:android="http://schemas.android.com/apk/res/android"
    xmlns:app="http://schemas.android.com/apk/res-auto"
```

```
    xmlns:tools="http://schemas.android.com/tools"
    tools:context="barry.demo.multilanguage.MainActivity" >
    <item
        android:id="@+id/action_english"
        android:orderInCategory="100"
        android:title="@string/english"
        android:showAsAction="never"
        />
    <item
        android:id="@+id/action_simple_chinses"
        android:orderInCategory="100"
        android:title="@string/simple_chinese"
        android:showAsAction="never"
        />
    <item
        android:id="@+id/action_traditional_chinese"
        android:orderInCategory="100"
        android:title="@string/traditional_chinese"
        android:showAsAction="never"
        />
</menu>
```

第九步：点击系统按键，实现标题中英文切换。具体如代码 CORE0410 所示。

代码 CORE0410：点击系统按键

```
/@Override
    public boolean onCreateOptionsMenu(Menu menu) {
            getMenuInflater().inflate(R.menu.main, menu);
            return true;
    }
    @Override
    public boolean onOptionsItemSelected(MenuItem item) {
            Configuration config = getResources().getConfiguration();
            switch (item.getItemId()) {
            case R.id.action_english:
                    config.locale = Locale.ENGLISH;
                    break;
            case R.id.action_simple_chinses:
```

```
                    config.locale = Locale.SIMPLIFIED_CHINESE;
                    break;
                case R.id.action_traditional_chinese:
                    config.locale = Locale.TRADITIONAL_CHINESE;
                    break;
                default:
                    return true;
                }
                getResources().updateConfiguration(config,    getResources().getDisplay-
Metrics());
                ((TextView)findViewById(R.id.textView1)).setText(R.string.hello_wor
ld);;
                return true;
        }
```

第十步：运行程序，运行结果如图 4.5 所示。

图 4.5 天津美景系统运行结果图

本项目主要介绍了关于 Android 应用资源的一些知识和运用，如语言国际化和 UI 界面颜色尺寸的应用变化，能够让我们更好地去开发一个应用程序。通过项目的学习，重点掌握如何

使用应用资源实现所需效果。

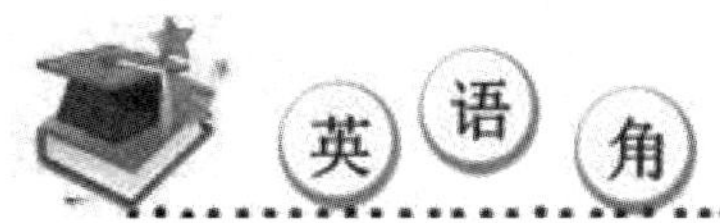

image	影像
view	视图
list	清单
array	数组
values	价值
simple	简单的
locale	场所地点
adapter	适配器
resources	来源
item	项目

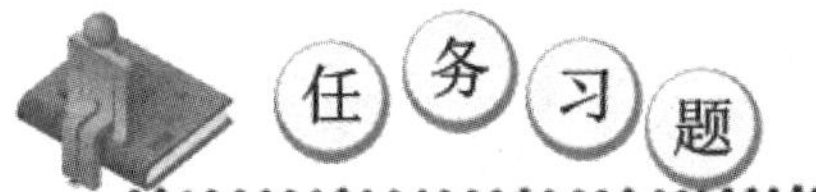

一、选择题

1. 下列(　　)项不是颜色定义的形式。

A.#RGB　　B.#ARGB　　C.#RRGGBB　　D.RRPCC

2. 下列(　　)项是中国语言代码的简写。

A.zh-cn　　B.en-za　　C.ko-kr　　D.pt-pt

3. 下列(　　)项是样式的标签。

A.<style>　　B.<theme>　　C. 与 <action>　　D.<service>

4. 下列(　　)项是完成国际化之后获得的新目录。

A.SDK 目录　　B.values-zh-rCN 目录　　C.Versona 目录　　D.android-ya 目录

5. Android 项目工程下面的 assets 目录的作用是(　　)。

A. 放置应用到的图片资源，res/drawable

B. 主要放置一些文件资源，这些文件会被原封不动打包到 apk 里面

C. 放置字符串，颜色，数组等常量数据 res/values

D. 放置一些与 UI 相应的布局文件，都是 xml 文件 res/layout

二、填空题

1. 国际化简称：______________，因为 i 和 n 之间还有 ________ 个字符，本地化，简称：______________。

2. 在 ____________，中编写界面所需颜色。

3. /res/drawable 下一般放置 ____________。

4. 控制控件的宽度的代码是 ____________。

5. Android 中动画一般 ____________ 目录下。

三、判断题

1. getResources().getString(R.string.hello) 可以得到一个 String 字符串。 （ ）

2. dp 是指屏幕像素。 （ ）

3. android:textSize="@dimen/some_name" 语法是否正确。 （ ）

4. 利用好 style 样式，可以减少我们的代码量。 （ ）

5. 我们在使用 UI 界面，并对其进行布局的时候，可以通过“android:textColor”对控件的文字惊醒颜色设置。 （ ）

四、简答题

1. 简要说一说 res 下的文件种类，并进行描述。

2. 说说 layout 的一些常见属性。

五、上机题

根据所学的知识自己设计一个美观的 UI 登录界面。

项目五　数据持久化操作

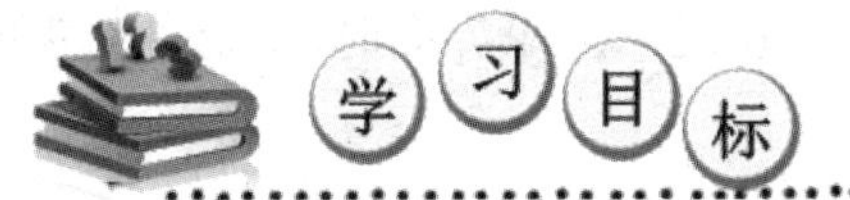

通过登录系统数据持久化操作，学习 Android 常用读写存储的相关知识，了解 SharedPreferences、SD 卡读写和 SQLite 存储的使用。在任务实现过程中：

- 掌握使用 SharedPreferences 读写移动智能系统的配置文件；
- 掌握读写 SD 卡的图片；
- 掌握使用 Android API 操作 SQLite 数据库。

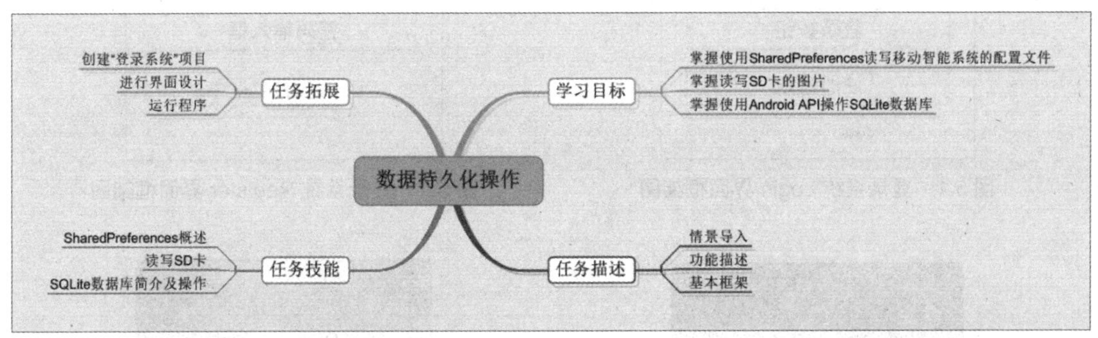

【情景导入】

随着智能手机的普及，越来越多的人习惯将一些重要资料保存到手机中，一旦手机没电关机，保存的数据也会随之丢失。因此 Android 提供了 SharedPreferences、SD 存储、SQLite 三种数据存储机制来保存数据。本次任务主要实现登录系统数据持久化操作。

【功能描述】

本任务将设计一款利用 SharedPreferences、SD 卡和 SQLite 存储的"登录系统"软件。

- 使用线性布局技术来设计登录系统界面；
- Logo 界面，Toast 提示“欢迎进入登录系统”；
- 登录界面，点击“注册”按钮，跳转到注册界面；
- 注册界面，输入姓名、性别、用户名、用户密码，点击图片拍照，点击“注册”按钮进行注册；
- 登录界面，输入已注册的用户名和用户密码，点击“登录”按钮进行登录；
- 登录界面，点击“记住密码”选择框，程序退出后再次进入登录界面时，用户名，用户密码依旧显示；
- 登录界面，点击“自动登录”选择框，程序退出后再次进入登录界面时，系统自动实现登录功能。

【基本框架】

基本框架如图 5.1、图 5.2 所示，将框架图转换成的效果如图 5.3 至图 5.6 所示。

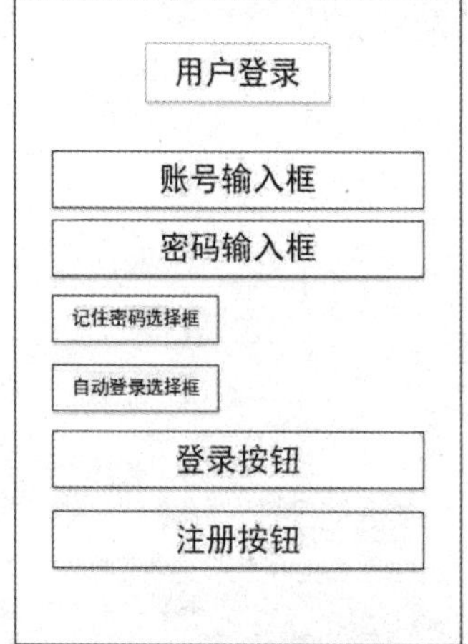

图 5.1 登录系统 Login 界面框架图

图 5.2 登录系统 Register 界面框架图

图 5.3 登录系统 Logo 界面效果图

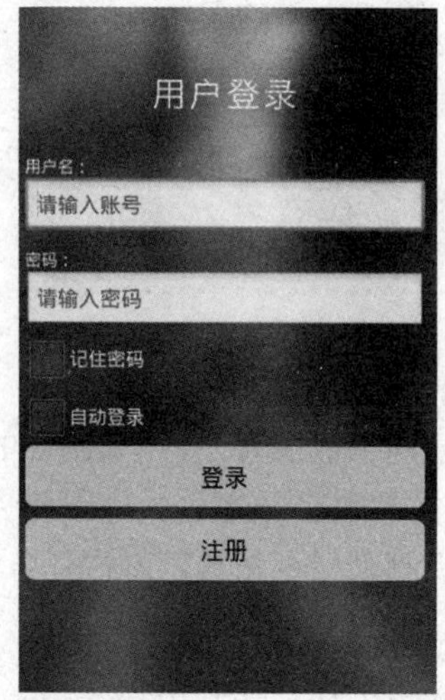

图 5.4 登录系统 Login 界面效果图

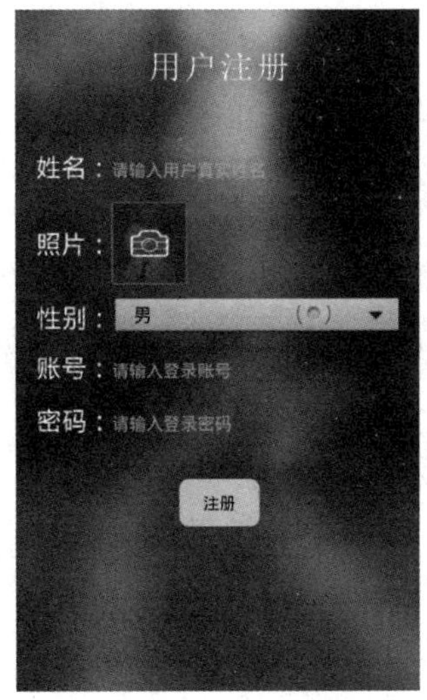

图 5.5　登录系统 Register 界面效果图

图 5.6　登录系统 Success 界面效果图

技能点 1　SharedPreferences 概述

1　SharedPreferences 的简介

SharedPreferences 存储类是以 XML 方式来保存，整体效率比较低，但对于常规的轻量级而言效率要高很多，如果存储量小可以考虑自己定义文件格式。XML 处理文件时是使用 Dalvik 通过自带底层的本地 XML Parser 解析的，XML Pull 方式也是使用的上述方法进行解析文件，这样对于内存资源占用比较好。

2　SharedPreferences 的特点

SharedPreferences 是 Android 平台上一个轻量级的存储类，有以下几个特点：

- 保存应用的一些常用配置。在 Activity 生命周期中了解当 Activity 执行 onpause() 时，最好存储数据，一般将此 Activity 的状态保存到 SharedPreferences 中。当 Activity 重载的时候，系统回调方法 onSaveInstanceState()，就能从 SharedPreferences 中将值取出。
- SharePreferences 提供了多种类型数据的保存接口，比如 long、int、String、char 类型接口。
- 可以全局共享访问。

3 SharedPreferences 操作模式

移动应用程序存储配置数据有四种模式，在上下文创建 SharedPreferences 实例对象的时候要指定目标访问应用程序的访问模式。四种操作模式如表 5.1 所示。

表 5.1 SharedPreferences 操作模式

操作模式	说明	值
Context.MODE_PRIVATE	默认操作模式，只能被应用本身访问	0
Context.MODE_APPEND	该模式会检查文件是否存在，存在就往文件追加内容，否则就创建新文件	32768
Context.MODE_WORLD_READABLE	表示当前文件可以被其他应用读取	1
Context.MODE_WORLD_WRITEABLE	表示当前文件可以被其他应用写入	2

4 SharedPreferences 常用方法及实现步骤

SharedPreferences 常用来存储一些轻量级的数据，常用的方法如表 5.2 所示。

表 5.2 SharedPreferences 常用方法

方法名称	含义
boolean getBoolean (String key, boolean defValue)	获取一个 boolean 类型的值
float getFloat (String key, float defValue)	获取一个 float 类型的值
int getInt (String key, int defValue)	获取一个 int 类型的值
long getLong (String key, long defValue)	获取一个 long 类型的值
String getString (String key, String defValue)	获取一个 String 类型的值
SharedPreferences.Editor edit ()	获取用于修改 SharedPreferences 对象设定值的接口引用
SharedPreferences.Editor putBoolean (String key, boolean value)	存入指定 key 对应的 boolean 值
SharedPreferences.Editor putFloat(String key, float value)	存入指定 key 对应的 float 值
SharedPreferences.Editor putInt(String key, int value)	存入指定 key 对应的 int 值
SharedPreferences.Editor putLong(String key, long value)	存入指定 key 对应的 long 值
SharedPreferences.Editor putString(String key, String value)	存入指定 key 对应的 String 值
SharedPreferences.Editor commit()	提交存入的数据

SharedPreferences 存储类在读取配置文件数据时，使用 getxx(key, defValue) 方法可获取配置文件中的数据。其中 key 代表 key-value 中的 key 值，defValue 代表如果配置文件中不存在此 key-value 键值配对时，则使用 defValue 代表默认值，保证程序正常运行。

SharedPreferences.Editor 工具接口提供了写入 XML 的方法，可以把指定数据类型以 key-value 的形式写入文件中。最后切记调用 commit() 方法提交存入的信息。

使用 SharedPreferences 存取 key-value 值一般需要经过以下步骤可以实现。

（1）通过 Context 上下文获取 SharedPreferences 实例对象 mSharedPreferences。

```
mSharedPreferences = context.getSharedPreferences(String name, int mode);
```

（2）sharedPreferences 对象取值或 sharedPreferences 对象获取 Edit 对象存值。

```
String temp = mSharedPreferences.getString(" 账号 ","123");// 获取账号
Editor mEditor = mSharedPreferences.edit();// 获取存值的工具对象
```

（3）通过 mEditor 对象存储 key-value 形式的配置数据。

```
mEditor.putString(" 账号 ",num);
```

（4）通过 mEditor 的 commit() 方法提交数据。

```
mEditor.commit();// 提交修改
```

技能点 2　读写 SD 卡

1　Environment 类的常用方法

移动设备的拍照频率高且每个图片都需要一定的存储空间，因此采用读写 SD 卡中的数据可以满足业务需求的存储要求。读写 SD 卡上的图片文件都是通过流的方式进行读取的，可以使用在 Java 面向对象程序设计中的流操作类。在读写 SD 卡时会常用到设备环境 android.os.Environment 工具类。Environment 类的常用方法说明如表 5.3 所示。

表 5.3　Environment 类常用方法

方法名称	含义
getDataDirectory()	获取 Android 中的 data 目录
getExternalStorgeDirectory()	获取到外部存储的目录，一般指 SD 卡
getDownloadCacheDirectory()	获取到下载的缓存目录
getExternalStorageState()	获取外部设置的当前状态
getRootDirectory()	获取到 Android Root 路径
isExternalStorageEmulated()	返回 Boolean 值判断外部设置是否有效

2 读写 SD 卡的文件的步骤

读写 SD 卡的文件的一般步骤为：

（1）判断移动设备中是否存在 SD 卡，如果存在则对 Android 系统 SD 卡里的文件操作添加使用权限。程序通过调用 Environment.getExternalStorageState() 方法的返回值与 Environment.MEDIA_MOUNTED 比较，如果 SD 卡存在并且具有操作权限则返回 true。Environment.getExternalStorageState().equals(android.os.Environment.MEDIA_MOUNTED)Android 中 SD 卡外部设置的状态情况如表 5.4 所示。

表 5.4 SD 卡状态说明

属性	含义
MEDIA_MOUNTED	可以进行读写
MEDIA_MOUNTED_READ_ONLY	存在，只可以进行读的操作

（2）通过调用 Environment.getExternalStorageDirectory() 获取文件绝对路径（即 /mnt/sdcard/+ 文件名），也可以在程序中直接写“ /mnt/sdcard/+ 文件名”这个字符串。

（3）获取文件路径后的操作，使用 FileInputStream、FileOutputStream、FileReader、FileWriter 四个类的方法实现读写 SD 卡文件数据。

（4）如果是在模拟器中测试使用 SD 卡，需要设置 SD 卡的大小如图 5.7 所示。在 Android 操作系统中操作 SD 卡需要在 AndroidManifest.xml 配置文件的 Permission 选项卡添加两个用户访问权限，如图 5.8 所示。

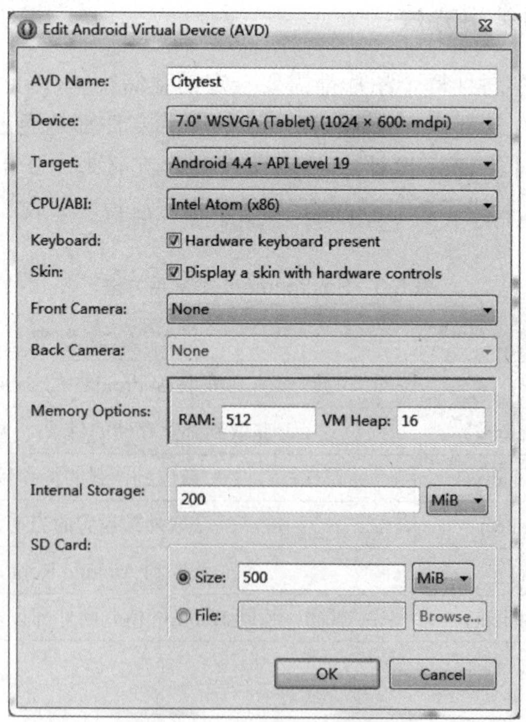

图 5.7 SD 卡 size 配置

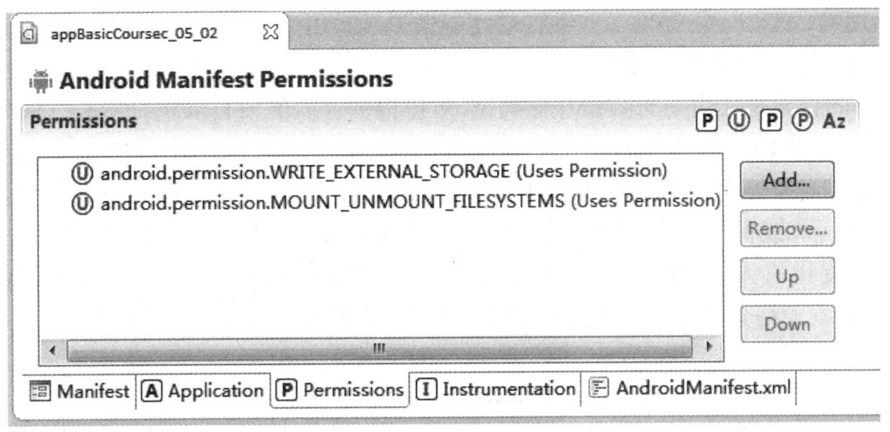

图 5.8 权限配置

（5）在 AndroidManifest.xml 配置文件的 Permission 选项卡中编辑 SD 的删除操作权限。

```
<uses-permission android:name="android.permission.MOUNT_UNMOUNT_FILESYSTEMS"/>
```

（6）在 AndroidManifest.xml 配置文件的 Permission 选项卡中编辑 SD 的写入操作权限。

```
<uses-permission android:name="android.permission.WRITE_EXTERNAL_STORAGE" />
```

技能点 3 SQLite 数据库简介及操作

1 SQLite 数据库简介

SQLite 轻量级关系型数据库发布于 2000 年。尽管 SQLite 是一个轻量级的数据库，但它支持关系型数据库（如 SQL Server、Oracle）操作数据的大部分功能，如触发器、索引、自动增长字段和 LIMIT/OFFSET 子句等。SQLite 数据库在运行时占用的系统资源极少，目前广泛地应用在嵌入式产品中。Android 平台已经嵌入了 SQLite 数据库，其具有如下特点。

- 跨平台：SQLite 可以编译运行在绝大多数主流操作平台上的软件，也适用于移动终端平台。
- 紧凑性：SQLite 一个功能齐全但体积很小数据库，可以描述为 1 个头文件，1 个库
- 适应性：作为一个内嵌式的数据库，具备强有力而且可伸缩的关系型数据库前端，简单而紧凑的多路搜索树后端。
- 不受拘束的授权：SQLite 的全部代码都在公共域中，不需要授权。
- 可靠性：SQLite 是一个开源的数据库，包含大约 30000 行标准 C 代码。

● 易用性：SQLite 还提供一些独特的功能来提高易用性，包括动态类型、冲突解决和“附加”多个数据库到一个连接的能力。

SQLite 的存储是采用动态数据存储类型，可以根据存入的值自动进行判断。SQLite 支持 5 种数据类型：NULL- 空值、INTEGER- 带符号的整型、REAL- 浮点型、TEXT- 字符串文本和 BLOB- 二进制对象。在实际编程过程中，SQLite 数据库可以写入 int、varchar 等大多数数据类型，在数据库运算或保存时将其转化为可以接受的 5 种数据类型。移动应用开发阶段只需要引用 SQLite 提供的 API 接口工具类即可创建和使用指定的数据库。通常在移动平台上使用 SQLiteDatabase 工具类创建或打开数据库的方法说明，如表 5.5 所示。

表 5.5 获取 SQLiteDatabase 数据库说明

<table>
<tr><th>方法</th><th>操作</th><th>参数说明</th></tr>
<tr><td>openDatabase(String path, SQLiteDatabase.CursorFactory factory, int flags, DatabaseErrorHandler errorHandler)</td><td>打开</td><td rowspan="5">path：数据库的路径；
factory：用于存储查询数据库的 Cursor 工厂，null 代表默认工厂；
flags：设置数据库访问模式（0：读写，1：只读）；
errorHandler：数据库处理异常的报告；
file：数据库文件；</td></tr>
<tr><td>openDatabase(String path, SQLiteDatabase.CursorFactory factory, int flags)</td><td>打开</td></tr>
<tr><td>openOrCreateDatabase(String path, SQLiteDatabase.CursorFactory factory, DatabaseErrorHandler errorHandler)</td><td>打开
创建</td></tr>
<tr><td>openOrCreateDatabase(String path, SQLiteDatabase.CursorFactory factory)</td><td>打开
创建</td></tr>
<tr><td>openOrCreateDatabase(File file, SQLiteDatabase.CursorFactory factory)</td><td>打开
创建</td></tr>
</table>

2 SQLite 数据库操作

基于数据库创建的基础上，可以对数据库进行建表并且对数据库表的 DML（数据操纵）进行操作。根据 Android API 中 SQLiteDatabase 提供的数据操作方法，下面通过表 5.6 列出常用的数据操作方法。

表 5.6 数据操作常用方法

方法	含义
execSQL(String sql)	执行标准 SQL 语句
execSQL(String sql, Object[] bindArgs)	执行带占位符的 SQL 语句
insert(String table, String nullColumnHack, ContentValues values)	插入一条数据
update(String table, ContentValues values, String whereClause, String[] whereArgs)	更新一条数据
delete(String table, String whereClause, String[] whereArgs)	删除一条数据

续表

方法	含义
beginTransaction()	开始事务
endTransaction()	结束事务

对数据库的一般操作步骤为：

（1）初始化打开或创建数据库：

```
SQLiteDatabase.openOrCreateDatabase(String path,SQLiteDatabase.CursorFactory factory));
```

（2）创建数据库表结构：

```
SQLiteDatabase.execSQL(String sql);
```

（3）执行数据操作：

```
SQLiteDatabase.insert(String table, String nullColumnHack, ContentValues values);
SQLiteDatabase.update(String table, ContentValues values, String whereClause, String[] whereArgs;
SQLiteDatabase.delete(String table, String whereClause, String[] whereArgs;
```

（4）关闭数据库：

```
SQLiteDatabase.close();
```

拓展：想了解或学习 SQLiteOpenHelper 工具类使用方法，可扫描下方二维码，获取更多信息。

第一步：在 Eclipse 中创建一个 Android 工程，命名为“登录系统”，并设计界面。如图 5.3 至图 5.6 所示。

第二步：在 src 文件夹下建立 LogoActivity.java 文件和 LoginActivity.java 文件，并实现从

闪屏界面跳转到登录界面。具体如代码 CORE0501 所示。

代码 CORE0501：LogoActivity 跳转

```
/**
        *1 此处完成界面延时跳转
         */
Toast.makeText(getApplicationContext(), " 欢迎进入登录系统 ",
                Toast.LENGTH_LONG).show();
        Thread splashTimer = new Thread() {
            public void run() {
                try {
                    long ms = 0;
                    while (m_bSplashActive && ms < m_dwSplashTime) {
                        sleep(30);
                        if (!m_bPaused)
                            ms += 30;
                    }
                startActivity(new android.content.Intent("MainActivity"));
                } catch (Exception ex) {
                    Log.e("Splash", ex.getMessage());
                } finally {
                    Intent intent = new Intent(LogoActivity.this,
                            LoginActivity.class);
                    startActivity(intent);
                    finish();
                }
            }
        };
        splashTimer.start();
    }
    protected void onPause() {
        super.onPause();
        m_bPaused = true;
    }
    protected void onResume() {
        super.onResume();
        m_bPaused = false;
        }
```

第三步：实现 Login 界面初始化，具体如代码 CORE0502 所示。

代码 CORE0502：实现 Login 界面初始化

```
// 初始化界面
edt_username = (EditText) findViewById(R.id.edt_username);
edt_userpsw = (EditText) findViewById(R.id.edt_userpsw);
remember = (CheckBox) findViewById(R.id.remember);
autologin = (CheckBox) findViewById(R.id.autologin);
```

第四步：在 src 文件夹下建立 RegisterActivity.java 文件并设置 Login 界面“注册”按钮事件，点击“注册”按钮，跳转到 Register 界面。具体如代码 CORE0503 所示。

代码 CORE0503：点击“注册”按钮跳转

```
// 跳转到注册页面进行注册
        startActivity(new Intent(LoginActivity.this, RegisterActivity.class));
```

第五步：在 src 文件夹下建立数据库 register.db 并建表 msg。具体如代码 CORE0504 所示。

代码 CORE0504：建立数据库表

```
/**
    *2 此处完成数据库及数据库表的创建
     */
public class registerHelper extends SQLiteOpenHelper {
    public registerHelper(Context context, String name, CursorFactory factory,
                int version) {
        super(context, name, factory, version);
        // TODO Auto-generated constructor stub
    }
    public registerHelper(Context context) {
        super(context, "register.db", null, 1);
        // TODO Auto-generated constructor stub
    }
    @Override
    public void onCreate(SQLiteDatabase arg0) {
        // TODO Auto-generated method stub
        // 创建数据库表名称为 msg
        arg0.execSQL("create table msg (Name text,Photo text,Sex text,Dname tex-
t,Psw text)");
```

```
        }
        @Override
        public void onUpgrade(SQLiteDatabase arg0, int arg1, int arg2) {
            // TODO Auto-generated method stub
        }
}
```

第六步：在 RegisterActivity.java 界面实现界面初始化，具体如代码 CORE0505 所示。

代码 CORE0505：实现 RegisterActivity.java 界面初始化

```
// 界面初始化
edt_psw = (EditText) findViewById(R.id.edt_psw);
sp_sex = (Spinner) findViewById(R.id.sp_sex);
edt_dname = (EditText) findViewById(R.id.edt_dname);
edt_name = (EditText) findViewById(R.id.edt_name);
img_photo = (ImageView) findViewById(R.id.img_photo);
btn_reg = (Button) findViewById(R.id.btn_reg);
```

第七步：在 RegisterActivity.java 界面输入姓名、账号、密码、性别，利用系统照相机拍照，添加注册信息。具体如代码 CORE0506 所示。

代码 CORE0506：添加注册信息

```
/**
    *3 此处完成注册信息添加
     */
str_photo = img_photo.toString();
        img_photo.setOnClickListener(new OnClickListener() {
            @Override
            public void onClick(View arg0) {
                // TODO Auto-generated method stub
                take_photo();
            }
        });
        // 给 Spinner 设置适配器
        String[] str = new String[] { " 男 ", " 女 " };
        ArrayAdapter<String> adapter = new ArrayAdapter<String>(this,
                android.R.layout.simple_spinner_dropdown_item, str);
        sp_sex.setAdapter(adapter);
        sp_sex.setOnItemSelectedListener(new OnItemSelectedListener() {
```

```
                @Override
                public void onItemSelected(AdapterView<?> arg0, View arg1,
                        int arg2, long arg3) {
                    // TODO Auto-generated method stub
                    // 根据索引判断选择的性别是男还是女。 0:男 1:女
                    switch (arg2) {
                    case 0:
                        str_sex = " 男 ";
                        break;
                    case 1:
                        str_sex = " 女 ";
                        break;
                    default:
                        break;
                    }
                }
                @Override
                public void onNothingSelected(AdapterView<?> arg0) {

                    // TODO Auto-generated method stub
                }
            });
    protected void take_photo() {
            // TODO Auto-generated method stub
            // 打开系统照相机
            startActivityForResult(
                    new Intent("android.media.action.IMAGE_CAPTURE"), 1);
        }
        @Override
        protected void onActivityResult(int requestCode, int resultCode, Intent data) {
            // TODO Auto-generated method stub
            super.onActivityResult(requestCode, resultCode, data);
            Bitmap bitmap = (Bitmap) data.getExtras().get("data");
            img_photo.setImageBitmap(bitmap);
            // 设置照片存储路径
            File f = new File("/mnt/sdcard/image");
            f.mkdir();
```

```
            FileOutputStream stream;
            try {
                // 将照片转换成文本的形式
                icon_path = "/mnt/sdcard/image" + System.currentTimeMillis()
                        + ".png";
                stream = new FileOutputStream(icon_path);
                bitmap.compress(Bitmap.CompressFormat.JPEG, 90, stream);
                stream.close();
            } catch (Exception e) {
                // TODO: handle exception
            }
        }
```

第八步：在 Register 界面设置"注册"按钮点击事件，点击"注册"按钮将界面上的所有信息保存到数据库中 。具体如代码 CORE0507 所示。

代码 CORE0507：设置"注册"按钮事件

```
/**
    *4 此处完成注册功能
     */

btn_reg.setOnClickListener(new OnClickListener() {
            @Override
            public void onClick(View arg0) {
                // TODO Auto-generated method stub
                str_dname = edt_dname.getText().toString().trim();
                str_psw = edt_psw.getText().toString().trim();
                str_name = edt_name.getText().toString().trim();
                // 判断注册信息是否为空 如果没有则进行注册
                if (!str_name.equals("") && !str_dname.equals("")
                        && !str_psw.equals("")) {
                    reg.insert(str_name, icon_path, str_sex, str_dname, str_psw);
                    Toast.makeText(RegisterActivity.this, " 注册成功 ", 0).show();
                    reg.close();
                finish();
                } else {
                    Toast.makeText(RegisterActivity.this, " 请完善信息后再进行
注册 ", 0)
```

```
.show();
                    }
                }
            });
```

第九步：将注册信息写入数据库表 msg 中。具体如代码 CORE0508 所示。

代码 CORE0508：写入数据库表

```
/**
     *5 此处填写插入数据库功能
      */
public void insert(String name, String icon_path, String str_sex,
                    String dname, String psw) {
          // TODO Auto-generated method stub
//         插入数据
          String sql="insert into msg
values('"+name+"','"+icon_path+"','"+str_sex+"','"+dname+"','"+psw+"')";
          System.out.println(sql);
          getWritableDatabase().execSQL(sql);
     }
```

第十步：返回到 Login 界面设置"登录"按钮事件。点击"登录"按钮查询数据库信息，具体如代码 CORE0509 所示。

代码 CORE0509：登录查询

```
/**
     *6 此处完成登录查询
      */
public void myclick(View v) {
          switch (v.getId()) {
          case R.id.btn_login:
          // 判断用户名密码是否正确
               getMsg();
               break;
          default:
               break;
          }
     }
private void getMsg() {
```

```
            // TODO Auto-generated method stub
            Cursor c=reg.equw();
            System.out.println(c.getCount()+"-----------------");
                if(c.moveToNext()){
                    userName=c.getString(3);
                    userpassword=c.getString(4);
                    System.out.println(" 账号: "+userName+" 密码: "+userpassword);
                }
                c.close();
                reg.close();
            }
```

第十一步：将查询信息语句写入数据库表 msg 中。具体如代码 CORE0510 所示。

代码 CORE0510：查询数据库信息

```
    /**
        *7 此处完成数据库查询
         */
    public Cursor equw(String username,String pwd) {
            // TODO Auto-generated method stub
            // 查询数据
            String sql="select * from msg where dname='"+username+"' and psw = '"+pw
d+"'";
            System.out.println(sql);
            Cursor c=getReadableDatabase().rawQuery(sql, null);
            return c;
    }
```

第十二步：在 Login 界面判断数据库信息与输入信息是否相同，根据判断结果进行登录或提示错误，并添加记住密码以及自动登录功能。具体如代码 CORE0511 所示。

代码 CORE0511：记住密码以及自动登录

```
    /**
        *8 此处完成记住密码功能及自动登录功能
         */
      p = getSharedPreferences("userInfo", 0);
            String name = sp.getString("USER_NAME", "");
            String pass = sp.getString("PASSWORD", "");
            boolean choseRemember = sp.getBoolean("remember", false);
```

```
            boolean choseAutoLogin = sp.getBoolean("autologin", false);
            // 如果上次选了记住密码,那进入登录页面也自动勾选记住密码,并填上
用户名和密码
            if (choseRemember) {
                edt_username.setText(name);
                edt_userpsw.setText(pass);
                remember.setChecked(true);
            }
            // 如果上次登录选了自动登录,那进入登录页面也自动勾选自动登录
            if (choseAutoLogin) {
                autologin.setChecked(true);
                getMsg();
                startActivity(new Intent(LoginActivity.this, SuccessActivity.class));

            }
            public void myclick(View v) {
            switch (v.getId()) {
            case R.id.btn_login:
            // 判断用户名密码是否正确
                d_userName = edt_username.getText().toString().trim();
                d_userpassword = edt_userpsw.getText().toString().trim();
                if(getMsg(d_userName, d_userpassword)){
                SharedPreferences.Editor editor = sp.edit();
                if (d_userName != null && d_userpassword != null) {
                    System.out.println(d_userName);
                        editor.putString("USER_NAME", d_userName);
                        editor.putString("PASSWORD", d_userpassword);
                        // 是否记住密码
                        if (remember.isChecked()) {
                            editor.putBoolean("remember", true);
                        } else {
                            editor.putBoolean("remember", false);
                        }
                        // 是否自动登录
                        if (autologin.isChecked()) {
                            editor.putBoolean("autologin", true);
                        } else {
                            editor.putBoolean("autologin", false);
```

```
                        }
                        editor.commit();
                        startActivity(new Intent(LoginActivity.this,
                                SuccessActivity.class));
                    } else {
                        Toast.makeText(LoginActivity.this, " 请输入用户名或密码 ", 0).
show();
                    }
                }
                else {
                    Toast.makeText(LoginActivity.this, " 用户名或密码不正确 ", 0).
show();
                }
                break;
            default:
                break;
            }
        }
        private Boolean getMsg(String username, String pwd) {
            // TODO Auto-generated method stub
            Cursor c = reg.equw(username, pwd);
            System.out.println(c.getCount() + "-----------------");
            if (c.getCount() > 0) {
                return true;
            } else {
                return false;
            }
        }
```

第十三步：运行程序，效果如图 5.3 至图 5.6 所示 。

【拓展目的】

熟悉并掌握数据持久化操作技能。

【拓展内容】

在“登录系统”基础上增加“忘记密码”功能。效果如图 5.9 至图 5.11 所示 。

图 5.9　Login 界面效果图

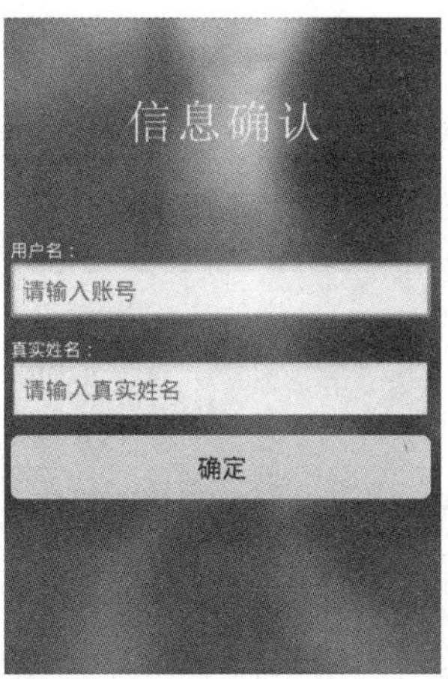

图 5.10　Forget 界面效果图

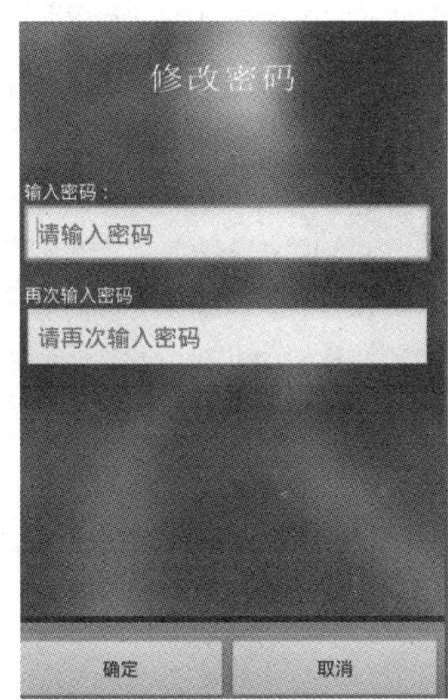

图 5.11　Update 界面效果图

【拓展步骤】

（1）设计思路：点击“忘记密码”后，用户可根据用户真实姓名与账号修改密码。

（2）修改信息判断以及提示信息，具体如代码 CORE0512 所示。

代码 CORE0512：判断信息并提示

```
        /**
      * 9 此处完成判断用户信息功能
        */
btn_fsure.setOnClickListener(new OnClickListener() {
            @Override
            public void onClick(View arg0) {
                // TODO Auto-generated method stub
```

```
                    final String str_name = edt_fname.getText().toString().trim();
                    final   String   str_username   =   edt_fusername.getText().toString().
trim();
                    //  判断数据库信息与输入信息是否相同
                    if(getMsg(str_name,str_username)){
                    // 编写 dialog 密码修改框
                        View view =
    LayoutInflater.from(ForgetActivity.this).inflate(R.layout.updata, null);
                        final EditText edt_sruepwd = (EditText) view.find
    ViewById(R.id.edt_surepwd);
                        final EditText edt_sruepwd1 = (EditText)
    view.findViewById(R.id.edt_surepwd1);
                        final AlertDialog.Builder builder=new
    AlertDialog.Builder(ForgetActivity.this);
                         AlertDialog dialog ;
                        builder.setIcon(android.R.drawable.ic_dialog_info);
                        builder.setView(view);

                        builder.setPositiveButton(" 确定 ", new
    DialogInterface.OnClickListener() {
                        @Override
                        public void onClick(DialogInterface arg0, int arg1) {
                            // TODO Auto-generated method stub
                        }
                    });
                     builder.setNegativeButton(" 取 消 ",      new    DialogInterface.
OnClickListener() {
                        @Override
                        public void onClick(DialogInterface arg0, int arg1) {
                            // TODO Auto-generated method stub
    }
                    });
                        AlertDialog dialog1 = builder.create();
                        dialog1.show();
                    }else{
                        Toast.makeText(ForgetActivity.this,  " 验 证 信 息 错 误 ",  0).
show();
                    }
```

```
            }
        });
```

（3）输入新密码进行密码修改，具体如代码 CORE0513 所示。

代码 CORE0513：密码修改

```
/**
     *10 此处完成密码修改
      */

builder.setPositiveButton(" 确定 ", new DialogInterface.OnClickListener() {
                        @Override
                        public void onClick(DialogInterface arg0, int arg1) {
                            // TODO Auto-generated method stub
                            String     str_sruepwd=edt_sruepwd.getText().toString().
trim();
                            String   str_sruepwd1=edt_sruepwd1.getText().toString().
trim();
                            if(!str_sruepwd.equals(str_sruepwd1)){
                                Toast.makeText(ForgetActivity.this, " 两次输入密
码不一致 ", 0).show();
                                Field field;
                            // 点击确定时 dialog 提示框显示
                                try {
                                    field =
 arg0.getClass().getSuperclass().getDeclaredField("mShowing");
                                    field.setAccessible(true);
                                        field.set(arg0, false);
                                } catch (Exception e) {
                                    // TODO Auto-generated catch block
                                    e.printStackTrace();
                                }
                                return;
                            }else{
                                db.update(str_sruepwd,str_name,str_username);
                                db.close();
                                try {
                                    Field field =
```

```
arg0.getClass().getSuperclass().getDeclaredField("mShowing");
                                    field.setAccessible(true);
                                    field.set(arg0, true);
                                } catch (Exception e) {
                                    // TODO Auto-generated catch block
                                    e.printStackTrace();
                                }
                            }
                        }
                    });
                    builder.setNegativeButton(" 取消 ", new DialogInterface.OnClick-
Listener() {
                        @Override
                        public void onClick(DialogInterface arg0, int arg1) {
                            // TODO Auto-generated method stub
                            // 点击取消时 dialog 显示框消失
                            try {
                                Field field =
arg0.getClass().getSuperclass().getDeclaredField("mShowing");
                                field.setAccessible(true);
                                field.set(arg0, true);
                            } catch (Exception e) {
                                // TODO Auto-generated catch block
                                e.printStackTrace();
                            }
                        }
                    });
```

(4)修改数据库数据代码，具体如代码 CORE0514 所示。

代码 CORE0514：修改数据库信息

```
/**
    *11 此处完成数据库修改
    */
public void update(String pwd,String name,String dname) {
        // TODO Auto-generated method stub
        String sql="update msg set psw='"+pwd+"' where name = '"+name+"' and
dname='"+dname+"'";
```

```
        System.out.println(sql);
        getWritableDatabase().execSQL(sql);
    }
```

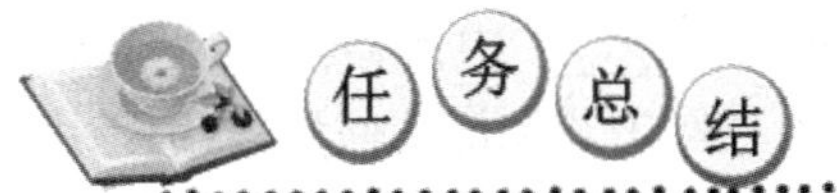

本项目内容主要介绍了在移动设备中应用存储相关技术的基础知识。重点讲解 SharedPreferences 存储配置参数、SD 卡的读写图片和 SQLite 数据库存储图书信息等技能。通过这些技能点在实际业务中的应用,明确在开发中采用合适的技术。

name	名字
password	密码
remember	记着
register	注册
sleep	睡眠
create	创造
insert	插入
update	更新
delete	删除

一、选择题

1. SharedPreferences 是一个(　　)的存储数据的方法。

A. 轻量级　　B. 较轻量级　　C. 重量级　　D. 较重量级

2. SQLite 是 Android 提供的一个(　　)数据库,并支持 SQL 语句。

A. 标准　　B. 小型　　C. 大型

3. ContentProvider 通过(　　)方法删除一行或多行数据。

A.delete()　　B.insert()　　C.query()　　D.update()

4. SQLiteOpenHelper 通过(　　)方法获取数据库名称。

A.close()　　B. getDatabaseName()　　C.getReadableDatabase()

5. Android 解析 XML 的方式有(　　)。(多选)

A. SAX　　B. DOM　　C. PULL

二、填空题

1. SharedPreferences：共享参数形式，一种以 ______ 的键值对形式保存数据的方式，Android 内置的，一般应用的配置信息，推荐使用此种方式保存。

2. SQLite Databases 在私有的数据库中存储 ______。

3. 内部存储，在 Android 中，开发者可以直接使用设备的 ______ 中保存文件，默认情况下，以这种方式保存的数据是只能被当前程序访问，在其他程序中是无法访问到的，而当用户卸载该程序的时候，这些文件也会随之被删除。

4. Internal Storage 在设备存储空间中保存 _____。

5. Preferences 的根节点，相当于 Preferences 各种组件的 _____。

三、判断题

1. SharedPreferences 对象本身只能获取数据，并不支持数据的存储和修改。 （ ）

2. SharedPreferences 用来储存 key-value 形式的数据，只可以用来存储基本的数据类型。 （ ）

3. SQLite 中的“SQL”指的是“结构化查询语言”。 （ ）

4. 在 DBMS 中，最高级别的数据存储单位就是数据库本身，其中包含由行和列构成的数据表。 （ ）

5. 对于数据库来说，修改现存的数据也是写入操作的一种。 （ ）

四、简答题

1. 请谈下 Android 的数据存储方式。

2. Android 的解析 XML 方式有哪些。

五、上机题

使用 SharedPreferences 保存数据：

```
SharedPreferences settings = context.getSharedPreferences("CONFIG", 0);
            settings.edit()
            .putString("ServerisDefaultConnection", String.valueOf(isChecked))
            .commit();
```

获取数据：

```
SharedPreferences settings = getSharedPreferences("CONFIG", 0);
                    StringisDefaultConnection = settings.getString("ServerisDefa-
ultConnection", "");
    if(! "".equals(isDefaultConnection))
    {
    if(isDefaultConnection.equals("true"))
```

```
{
//do something
}
}
```

利用所给代码，写出保存并获取数据的程序。

项目六　复杂数据展示

通过“数据显示系统”的复杂数据显示，学习 Android 高级控件的相关知识，了解 Spinner、ListView、GridView 的使用方法。在任务实现过程中：

- 掌握下拉列表 Spinner 的功能与用法；
- 掌握 ListView 的功能与用法；
- 掌握 GridView 的功能与用法。

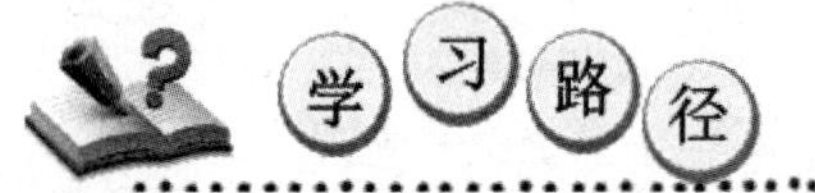

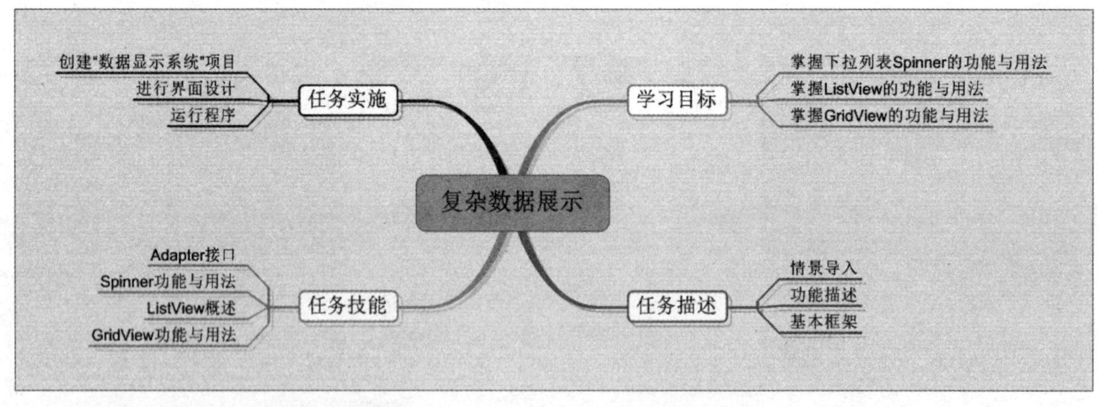

【情景导入】

智能手机已成为人们生活中必不可少的一部分，用户通过手机应用程序查看所需数据的情况也越来越多，如何高效率利用手机端查看用户所需要的信息，由此提出了基于手机端的复杂数据展示的开发来解决这个问题。本次任务主要实现复杂数据的展示。

【功能描述】

本项目将设计一款使用高级控件 Spinner、ListView、GridView 显示数据的软件。

- 使用线性布局技术来设计数据显示系统界面;
- 点击网格视图条目跳转到对应的人物信息列表;
- 点击“Spinner”选择条目,选择显示的样式。

【基本框架】

基本框架如图 6.1、图 6.2 所示,将框架图转换成的效果图如图 6.3、图 6.4 所示。

图 6.1　数据显示系统框架图

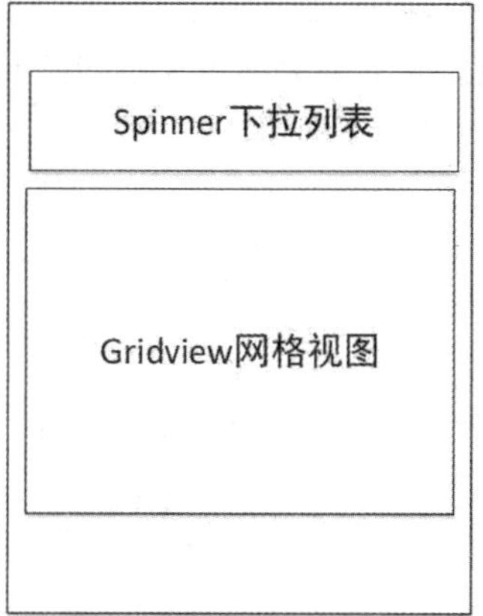

图 6.2　数据显示系统框架图

图 6.3　数据显示系统主界面效果图

图 6.4　数据显示系统列表效果图

技能点 1 Adapter 接口

Adapter 本身是一个接口，派生了两个子接口，分别是 ListAdapter 和 SpinnerAdapter。Adapter 接口及实现类继承关系如图 6.5 所示。

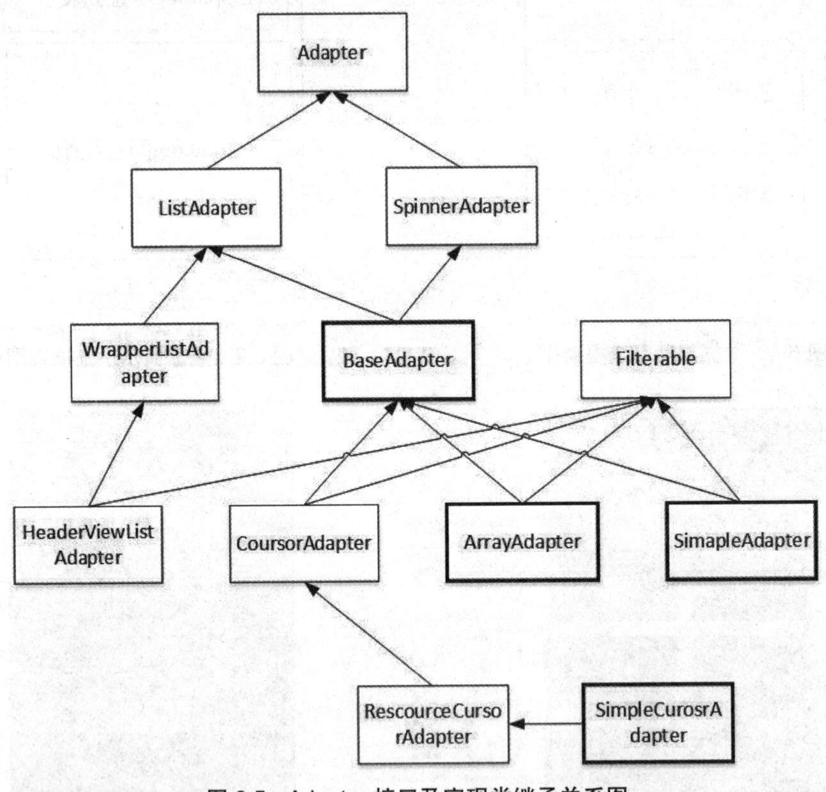

图 6.5 Adapter 接口及实现类继承关系图

从图 6.5 中可看出，几乎所有的 Adapter 都继承了 BaseAdapter，而 BaseAdapter 同时实现了 ListAdapter，SpinnerAdapter 接口。继承关系类图中粗线框表示常用 Adapter。

- SimpleAdapter：功能十分强大，可用于将 List 集合的多个对象包装成多个列表项。
- ArrayAdapter：简单好用，用于数组或 List 集合多值封装的多个列表项。
- BaseAdapter：一般用于被扩展，能够对各个列表项进行最大限度的定制。

拓展：想了解或学习适配器使用详情，可扫描下方二维码，获取更多信息。

技能点 2　Spinner 功能与用法

1　Spinner 控件简介

Spinner 是最常用的高级控件之一。当用户点击该控件时，会弹出选择列表供用户选择。选择列表中的元素都是来自适配器，每次屏幕只显示用户选中的元素。Spinner 提供了 UI 设计模式有更好的体验性。

2　Spinner 的常用属性及对应方法

Spinner 的常用属性及对应方法如表 6.1 所示，若开发者使用 Spinner 时已确定列表选择框里的列表项，则不需要编写代码，只要指定 android:entries 属性就可以正常工作。

表 6.1　Spinner 的常用属性及对应方法

属性名称	对应方法	属性说明
android:spinnerMode	setSprinnerMode(int)	设置 Spinner 样式，dropdown 和 dialog 两种
android:entries		使用数组资源设置 Spinner 下拉菜单的列表框
android:dropDownVerticalOffset	setDropDownVerticalOffset(int)	设置 Spinner 下拉菜单的水平偏移
android:dropDownHorizontalOffset	setDropDownHorizontalOffset(int)	设置 Spinner 下拉菜单的垂直偏移
android:dropDownWidth	setDropDownWidth(int)	设置 Spinner 下拉菜单的宽度
android:popupBackground	setPopupBackgroundResource()	设置下拉菜单的背景
android:prompt		设置 Spinner 下拉菜单的提示信息

技能点 3 ListView 概述

1 ListView 简介

ListView 是手机系统中使用非常广泛的一种组件，所有列表项都是以垂直的形式显示，生成列表视图有两种方式：

- 直接使用 ListView 进行创建。

```
ListView listView = new ListView(this);
```

- 让 Activity 继承 ListActivity。

```
public class ListView extends ListActivity {
public static void main(String[] args) {
    // TODO Auto-generated method stub
    }
}
```

2 ListView 属性

Android 开发中时常用到 ListView，它能够根据内容数据的长度自适应显示数据。可以引用 values 目录下的 array.xml 数组元素，也可以引用代码中自定义的数组元素，每一行数据为一条 Item。ListView 相关属性表如表 6.2 所示。

表 6.2 ListView 相关属性表

属性名称	属性说明
android:divder	每条 item 之间的分割线
android:divderHeight	分割线的高度
android:entries	引用一个将使用在此 ListView 里的数组，该数组定义在 value 目录下的 arrays.xml 文件中
android:footerDivdersEnabled	设为 flase 时，此 ListView 将不会在页脚视图前画分隔符，默认值为 true
android:headerDivdersEnabled	设为 flase 时，此 ListView 将不会在页眉视图前画分隔符，默认值为 true

技能点 4　GridView 功能与用法

1　GridView 控件的简介

GridView 与 ListView 可以对比来学习，这里重点讲 GridView。比如实现九宫格图，首选用 GridView，也是最简单的。GridView 与 ListView 相同，都需要通过 Adapter 来提供现实数据。开发人员可选用上面三种适配器进行信息填充。平时的手机页面就是用 GridView 做的，如图 6.6 所示。

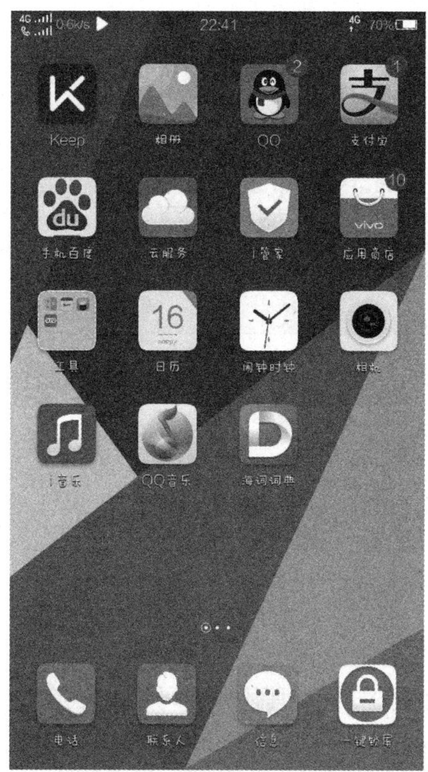

图 6.6　手机界面效果图

2　GridView 的属性

GridView 的列是用 numColumns 来指定的。GridView 一般设置 numColumns 大于 1，不设置时默认值为 1，如果设置为 3，则显示 3 列。GridView 提供的常用属性及相关方法如表 6.3 所示。

表 6.3 GridView 的属性

属性	解释
android:numColumns="auto_fit"	列数设置为自动
android:columnWidth="90dp"	每列的宽度，即 Item 的宽度
android:stretchMode="columnWidth	缩放与列宽大小相同
android:verticalSpacing="10dp"	垂直边距
android:horizontalSpacing="10dp"	水平边距

3 GridView 的用法

GridView 需要在 XML 文件里写出。GridView 的一些属性，具体实现方法如下所示：

```
<GridView
        android:id="@+id/gv_menu"
        android:layout_width="wrap_content"
        android:layout_height="wrap_content"
        android:layout_centerHorizontal="true"
        android:numColumns="2" >
   </GridView>
```

想要在程序中运用 GridView 控件，需要准备数据源、新建适配器、加载适配器。

第一步：在 Eclipse 中创建一个 Android 工程，命名为“数据显示系统”，并设计界面。如图 6.2 所示。

第二步：在 src 文件夹中建立 MainActivity.java 文件并实现界面初始化。

第三步：编写 GridView 适配器内容，具体如代码 CORE0601 所示。

代码 CORE0601：编写 GridView 适配器内容

```
/**
 * 1 此处填写 GridView 适配器
 */
public class GirdAdapter extends BaseAdapter {
     private int[] image = new int[] { R.drawable.no1, R.drawable.no2,
               R.drawable.no3, R.drawable.no4 };
     private Context mContext;
     public GirdAdapter(Context mContext) {
```

```
        // TODO Auto-generated constructor stub
        this.mContext = mContext;
    }
    @Override
    public int getCount() {
        // TODO Auto-generated method stub
        return image.length;
    }
    @Override
    public Object getItem(int arg0) {
        // TODO Auto-generated method stub
        return image[arg0];
    }
    @Override
    public long getItemId(int arg0) {
        // TODO Auto-generated method stub
        return arg0;
    }
    @Override
    public View getView(int arg0, View arg1, ViewGroup arg2) {
        // TODO Auto-generated method stub
        arg1 = LayoutInflater.from(mContext).inflate(R.layout.gird, null);
        ImageView ly = (ImageView) arg1.findViewById(R.id.lin);
        ly.setBackgroundResource(image[arg0]);
        return arg1;
    }
}
```

第四步：将 GridView 适配器填充到 GridView中，并设置 GridView 条目单击监听事件进行跳转，使用 Intent 传参。具体如代码 CORE0602 所示。

代码 CORE0602：适配器填充

```
/**
 * 2 此处填写填充适配器代码
 */
gv_menu.setAdapter(adapter);
gv_menu.setOnItemClickListener(new OnItemClickListener() {
            @Override
```

```
                public void onItemClick(AdapterView<?> arg0, View arg1, int arg2,
                        long arg3) {
                    // TODO Auto-generated method stub
                    Intent intent = new Intent();
                    switch (arg2) {
                    case 0:
                        Toast.makeText(MainActivity.this, " 王子信息 ", 0).show();
                        intent.setClass(MainActivity.this, UserActivity.class);
                            break;
                        case 1:
                            Toast.makeText(MainActivity.this, " 小 美 信 息 ", 0).
show();
                            intent.setClass(MainActivity.this, UserActivity.class);
                            break;
                        case 2:
                            Toast.makeText(MainActivity.this, " 大 熊 信 息 ", 0).
show();
                            intent.setClass(MainActivity.this, UserActivity.class);
                            break;
                        case 3:
                            Toast.makeText(MainActivity.this, " 壮 壮 信 息 ", 0).
show();
                            intent.setClass(MainActivity.this, UserActivity.class);
                            break;
                    default:
                        break;
                    }
                    intent.putExtra("type", arg2);
                    startActivity(intent);


                }
            });
```

第五步：编写 ListView 适配器内容，具体如代码 CORE0603 所示。

代码 CORE0603：ListView 适配器内容

```
/**
 * 3 此处填写 ListView 适配器代码
 */
public class ListAdapter extends BaseAdapter {
    String[] str_xmenu=new String[]{" 姓名 "," 性别 "," 身高 "," 体重 "," 爱好 "," 特长 "," 最喜欢的运动 "};
    String[] str_xmenu_text=new String[]{" 王子 "," 男 ","1.75CM","50KG"," 看书 "," 智力高 "," 跑步 "};
    String[] str_mmenu=new String[]{" 姓名 "," 性别 "," 身高 "," 体重 "," 爱好 "," 特长 "," 最喜欢的运动 "};
    String[] str_mmenu_text=new String[]{" 小美 "," 女 ","1.60CM","45KG"," 唱歌 "," 跳舞 "," 化妆 "};
    String[] str_fmenu=new String[]{" 姓名 "," 性别 "," 身高 "," 体重 "," 爱好 "," 特长 "," 最喜欢的运动 "};
    String[] str_fmenu_text=new String[]{" 大熊 "," 男 ","1.80CM","65KG"," 运动 "," 力气大 "," 跑步 "};
    String[] str_lmenu=new String[]{" 姓名 "," 性别 "," 身高 "," 体重 "," 爱好 "," 特长 "," 最喜欢的运动 "};
    String[] str_lmenu_text=new String[]{" 壮壮 "," 男 ","1.65CM","50KG"," 睡觉 "," 吃东西 "," 睡觉，睡觉，睡觉 "};
    Context context;
    int type;
    int sp_type;
    private LinearLayout ly_lin;
    public ListAdapter(Context context, int type,int sp_type) {
        // TODO Auto-generated constructor stub
        this.context=context;
        this.type=type;
        this.sp_type=sp_type;
    @Override
    public int getCount() {
        // TODO Auto-generated method stub
        return str_xmenu.length;
    }
    @Override
    public Object getItem(int arg0) {
        // TODO Auto-generated method stub
```

```
        return str_xmenu[arg0];
    }
    @Override
    public long getItemId(int arg0) {
        // TODO Auto-generated method stub
        return arg0;
    }
    @SuppressLint("ResourceAsColor") @Override
    public View getView(int arg0, View arg1, ViewGroup arg2) {
        // TODO Auto-generated method stub
        arg1 = LayoutInflater.from(context).inflate(R.layout.list, null);
        TextView tv_menu = (TextView) arg1.findViewById(R.id.tv_menu);
        TextView tv_text = (TextView) arg1.findViewById(R.id.tv_text);
        ly_lin = (LinearLayout) arg1.findViewById(R.id.ly_lin);
        switch (type) {
        case 0:
            tv_menu.setText(str_xmenu[arg0]);
            tv_text.setText(str_xmenu_text[arg0]);
            ly_lin.setBackgroundColor(Color.parseColor("#66b3ff"));
            break;
        case 1:
            tv_menu.setText(str_mmenu[arg0]);
            tv_text.setText(str_mmenu_text[arg0]);
            ly_lin.setBackgroundColor(Color.parseColor("#ff79bc"));
            break;
        case 2:
            tv_menu.setText(str_fmenu[arg0]);
            tv_text.setText(str_fmenu_text[arg0]);
                        ly_lin.setBackgroundColor(Color.parseColor("#ffbb77"));
            break;
        case 3:
            tv_menu.setText(str_lmenu[arg0]);
            tv_text.setText(str_lmenu_text[arg0]);
            ly_lin.setBackgroundColor(Color.parseColor("#ffff6f"));
            break;
        default:
            break;
        }
```

```
            if(sp_type==1){
                System.out.println("#1281f0");
                ly_lin.setBackgroundColor(Color.parseColor("#66b3ff"));
            }else if(sp_type==2){
                ly_lin.setBackgroundColor(Color.parseColor("#ff79bc"));
            }
            else if(sp_type==3){
                ly_lin.setBackgroundColor(Color.parseColor("#ffbb77"));
            }
            else if(sp_type==4){
                ly_lin.setBackgroundColor(Color.parseColor("#ffff6f"));
            }
            return arg1;
        }
    }
```

第六步：将 ListView 适配器填充到 ListView 中，并编写 Spinner 选择样式代码。具体如代码 CORE0604 所示。

代码 CORE0604：array.xml

```
    /**
     * 4 此处填写填充 Spinner 适配器代码
     */

    adapter = new ListAdapter(this, type,sp_type);
            lv_list.setAdapter(adapter);
            String[] str = new String[] { " 请选择样式 "," 主题样式 1", " 主题样式 2", "
主题样式 3", " 主题样式 4" };
            ArrayAdapter<String> sp_adapter = new ArrayAdapter<String>(this,
                    android.R.layout.simple_spinner_dropdown_item, str);
            sp_check.setAdapter(sp_adapter);
            sp_check.setSelection(0, true);
            sp_check.setOnItemSelectedListener(new OnItemSelectedListener() {
                @SuppressLint("ResourceAsColor")
                @Override
                public void onItemSelected(AdapterView<?> arg0, View arg1,
                        int arg2, long arg3) {
                    // TODO Auto-generated method stub
```

```
            switch (arg2) {
            case 0:
                arg2 = 5;
                break;
            case 1:
                System.out.println(" 主题样式 1：" +arg2 );
                break;
            case 2:
                System.out.println(" 主题样式 2：" + arg2);
                break;
            case 3:
                System.out.println(" 主题样式 3：" + arg2);
                break;
            case 4:
                System.out.println(" 主题样式 4：" + arg2);
                break;
            default:

                break;
            }
            sp_type=arg2;
            adapter = new ListAdapter(UserActivity.this, type, sp_type);
            System.out.println(sp_type+"--------------");
            lv_list.setAdapter(adapter);
        }
        @Override
        public void onNothingSelected(AdapterView<?> arg0) {
            // TODO Auto-generated method stub
        }
        });
```

第七步：运行程序，运行结果如图 6.3、图 6.4 所示。

【拓展目的】

熟悉并掌握使用高级控件显示复杂数据等技巧。

【拓展内容】

在“数据显示系统”基础上增加 Spinner 选择人员信息功能，选择效果如图 6.3、图 6.4、图 6.7 所示。

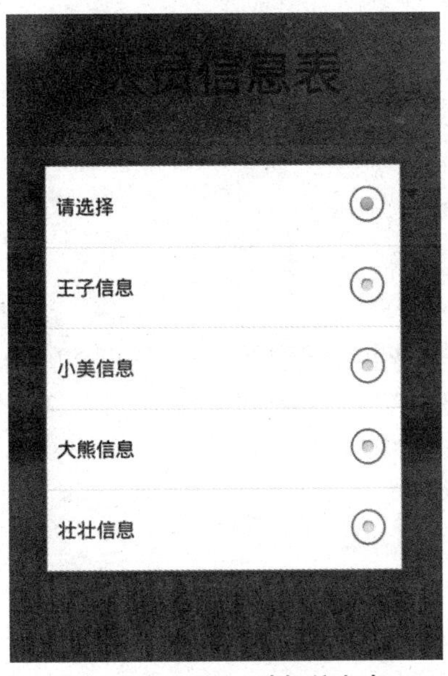

图 6.7 Spinner 选择信息表

【拓展步骤】

（1）修改适配器中的填充的数据，实现 Spinner 下拉列表选择人员信息功能。

（2）编写 ListView 适配器内容，具体如代码 CORE0605 所示。

代码 CORE0605：ListView 适配器内容

```
/**
 *  1 此处填写填充 ListView 适配器代码
 */
public class ListAdapter extends BaseAdapter {
    String[] str_xmenu=new String[]{" 姓名 "," 性别 "," 身高 "," 体重 "," 爱好 "," 特长 "," 最喜欢的运动 "};
    String[] str_xmenu_text=new String[]{" 王子 "," 男 ","1.75CM","50KG"," 看书 "," 智力高 "," 跑步 "};
    String[] str_mmenu=new String[]{" 姓名 "," 性别 "," 身高 "," 体重 "," 爱好 "," 特长 "," 最喜欢的运动 "};
    String[] str_mmenu_text=new String[]{" 小美 "," 女 ","1.60CM","45KG"," 唱歌 "," 跳舞 "," 化妆 "};
    String[] str_fmenu=new String[]{" 姓名 "," 性别 "," 身高 "," 体重 "," 爱好 "," 特长 "," 最喜欢的运动 "};
```

```
        String[] str_fmenu_text=new String[]{" 大熊 "," 男 ","1.80CM","65KG"," 运动
"," 力气大 "," 跑步 "};
        String[] str_lmenu=new String[]{" 姓名 "," 性别 "," 身高 "," 体重 "," 爱好 "," 特
长 "," 最喜欢的运动 "};
        String[] str_lmenu_text=new String[]{" 壮壮 "," 男 ","1.65CM","50KG"," 睡觉
"," 吃东西 "," 睡觉，睡觉，睡觉 "};
        Context context;
        int type;
        int sp_type;
        private LinearLayout ly_lin;
        public ListAdapter(Context context, int type,int sp_type) {
            // TODO Auto-generated constructor stub
            this.context=context;
            this.type=type;
            this.sp_type=sp_type;
        }
        @Override
        public int getCount() {
            // TODO Auto-generated method stub
            return str_xmenu.length;
        }

        @Override
        public Object getItem(int arg0) {
            // TODO Auto-generated method stub
            return str_xmenu[arg0];
        }
        @Override
        public long getItemId(int arg0) {
            // TODO Auto-generated method stub
            return arg0;
        @SuppressLint("ResourceAsColor") @Override
        public View getView(int arg0, View arg1, ViewGroup arg2) {
            // TODO Auto-generated method stub
            arg1 = LayoutInflater.from(context).inflate(R.layout.list, null);
            TextView tv_menu = (TextView) arg1.findViewById(R.id.tv_menu);
            TextView tv_text = (TextView) arg1.findViewById(R.id.tv_text);
            ly_lin = (LinearLayout) arg1.findViewById(R.id.ly_lin);
```

```
switch (type) {
case 0:
    tv_menu.setText(str_xmenu[arg0]);
    tv_text.setText(str_xmenu_text[arg0]);
    ly_lin.setBackgroundColor(Color.parseColor("#66b3ff"));
    break;
case 1:
    tv_menu.setText(str_mmenu[arg0]);
    tv_text.setText(str_mmenu_text[arg0]);
    ly_lin.setBackgroundColor(Color.parseColor("#ff79bc"));
    break;
case 2:
    tv_menu.setText(str_fmenu[arg0]);
    tv_text.setText(str_fmenu_text[arg0]);
    ly_lin.setBackgroundColor(Color.parseColor("#ffbb77"));
    break;
case 3:
    tv_menu.setText(str_lmenu[arg0]);
    tv_text.setText(str_lmenu_text[arg0]);
    ly_lin.setBackgroundColor(Color.parseColor("#ffff6f"));
    break;
default:
    break;
}
if(sp_type==1){
    System.out.println("#1281f0");
    tv_menu.setText(str_xmenu[arg0]);
    tv_text.setText(str_xmenu_text[arg0]);
    ly_lin.setBackgroundColor(Color.parseColor("#66b3ff"));
}else if(sp_type==2){
    tv_menu.setText(str_mmenu[arg0]);
    tv_text.setText(str_mmenu_text[arg0]);
    ly_lin.setBackgroundColor(Color.parseColor("#ff79bc"));
}
else if(sp_type==3){
    tv_menu.setText(str_fmenu[arg0]);
    tv_text.setText(str_fmenu_text[arg0]);
    ly_lin.setBackgroundColor(Color.parseColor("#ffbb77"));
```

```
            }
            else if(sp_type==4){
                tv_menu.setText(str_lmenu[arg0]);
                tv_text.setText(str_lmenu_text[arg0]);
                ly_lin.setBackgroundColor(Color.parseColor("#ffff6f"));
            }else {
            }
            return arg1;
        }
    }
```

（3）填充适配器代码，具体如代码 CORE0606 所示。

代码 CORE0606：ListView 单击条目事件

```
/**
 * 2 此处填写 ListView 单击条目事件
 */
adapter = new ListAdapter(this, type,sp_type);
        lv_list.setAdapter(adapter);
        String[] str = new String[] { " 请选择 "," 王子信息 "," 小美信息 "," 大熊信
息 "," 壮壮信息 " };
        ArrayAdapter<String> sp_adapter = new ArrayAdapter<String>(this,
                android.R.layout.simple_spinner_dropdown_item, str);
        sp_check.setAdapter(sp_adapter);
        sp_check.setSelection(0, true);
        sp_check.setOnItemSelectedListener(new OnItemSelectedListener() {
            @SuppressLint("ResourceAsColor")
            @Override
            public void onItemSelected(AdapterView<?> arg0, View arg1,
                    int arg2, long arg3) {
                // TODO Auto-generated method stub
                switch (arg2) {
                case 0:
                    break;
                case 1:
                    System.out.println(" 主题样式 1：" + sp_type);
                    break;
                case 2:
```

```
                System.out.println(" 主题样式 2：" + sp_type);
                break;
            case 3:
                System.out.println(" 主题样式 3：" + sp_type);
                break;
            case 4:
                System.out.println(" 主题样式 4：" + sp_type);
                break;
            default:
                break;
            }

            sp_type=arg2;
            adapter = new ListAdapter(UserActivity.this, type, sp_type);
            System.out.println(sp_type+"--------------");
            lv_list.setAdapter(adapter);
        }
        @Override
        public void onNothingSelected(AdapterView<?> arg0) {
            // TODO Auto-generated method stub
        }
        });
```

本项目主要介绍了 Android 平台中的一些复杂数据展示的控件，Spinner、ListView 和 GridView 在开发中较为常用。通过本项目的学习，读者需要熟练掌握控件及适配器使用方法，对复杂数据展示有进一步的了解。

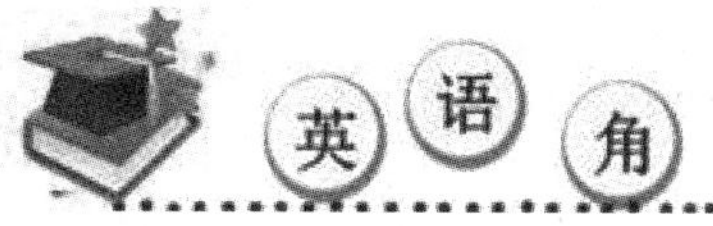

list	清单
entry	入口处
false	错误
String	字符串
grid	方格

break	中断
type	类型
base	基础
import	导入
switch	开关

一、选择题

1. Spinner 以下拉菜单形式显示时，它的控制属性是（　　）。

A. spinnerMode 属性　　B. entries 属性

C. dropDownSelector 属性　　D. dropDownWidth 属性

2. 下列关于 ListView 使用的描述中，不正确的是（　　）。

A. 要使用 ListView，必须为该 ListView 使用 Adapter 方式传递数据

B. 要使用 ListView，该布局文件对应的 Activity 必须继承 ListActivity

C. ListView 中大的每一个项的视图既可以使用内置的布局，也可以使用自定义的布局方式

D. ListView 中的每一个项被选中时，将会触发 ListView 对象的 itemclick 事件

3. 下列用以显示一系列图像的是（　　）。

A. ImageView　　B. Gallery　　C. ImageSwitcher　　D. GridView

4. 表示下拉列表的组件是（　　）。

A. Gallery　　B. Spinner　　C. GridView　　D. Listview

5. 在 Android 程序的调试过程中下列可以使用的方式不包括（　　）。

A. DDMS　　B. 使用 Java 程序在控制台输出

C. 删除错误代码　　D. 设置断点

二、填空题

1. ListView 的配适器有哪些？请列举 3 种：______、______、______。

2. 提高 ListView 的运行效率可以借助 ______ 来进行优化。

3. ArrayAdapter 中得到当前项的对象实例的方法是：______。

4. 手机屏幕上主界面的 App 图标列表使用 ______ 控件实现的。

5. 为 ListView 添加适配器的代码是：______。

三、判断题

1. Spinner 显示项的数据绝对不可能用 ArrayList。（　　）

2. Adapter 的作用是界面与数据之间的桥梁，通过设置适配器至 ListView 控件后（如调用 ListView 的 setAdapter(ListAdapter adapter)，列表的每一项会显示至页面中。其实，当列表里的每一项显示到页面时，都不会调用 Adapter 的 getView 方法返回一个 View。（　　）

3．android:gravity 是设置对齐方式的代码。　　(　　)

4. ListView 是手机系统中使用非常广泛的一种组件，它以垂直列表的形式显示所有列表项。　　(　　)

5. android:prompt 这个代码是设置下拉菜单的提示信息。　　(　　)

四、简答题

1. 简单描述一下 GridView 控件与 ListView 控件区别。

2. 列举 3 个 ListView 的适配器，并加以描述。

五、上机题

ListView 支持多种显示方法，请用不同显示方法对应不同数据描述集合。如下图手机界面所示。

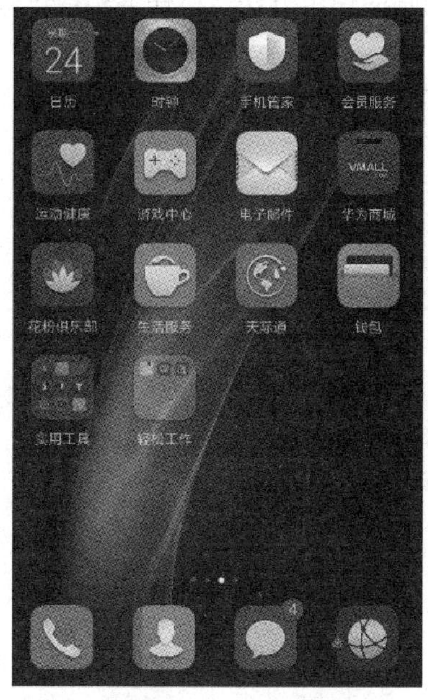

项目七　图形图像

通过实现图形图像的处理及动画功能，学习 Bitmap 位图的使用以及帧动画相关知识，了解逐帧动画、补间动画和属性动画的使用方法。在任务实现过程中：

- 掌握逐帧动画的使用方法；
- 掌握补间动画的使用方法；
- 掌握属性动画的使用方法。

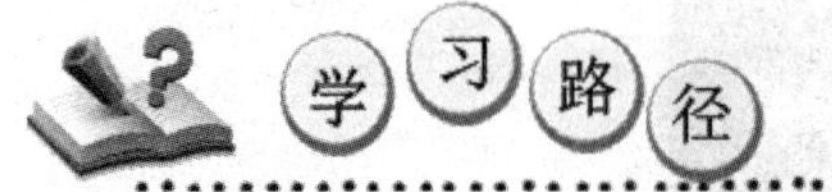

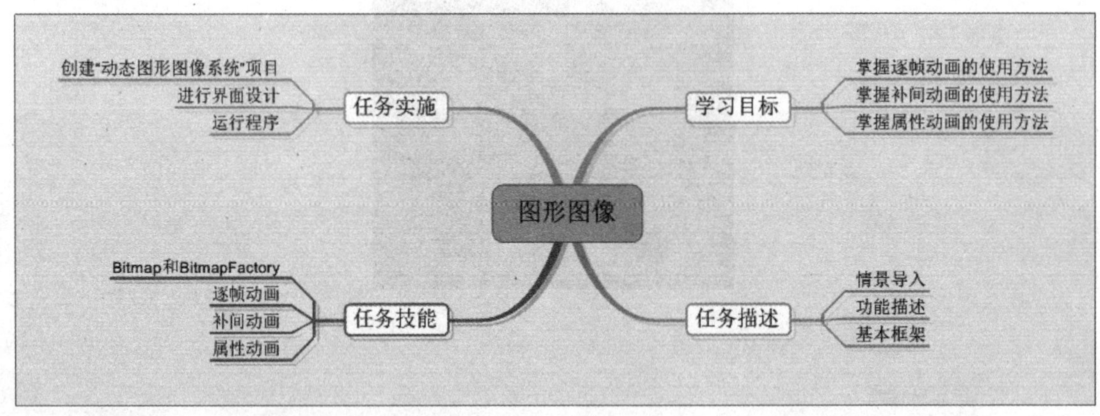

【情景导入】

随着 Android 技术的高速发展，图形图像效果的处理也得到了提升，为了能够成功实现用户所需要的图形图像的效果，需要对图形图像的处理进行进一步了解。本项目主要实现动态图形图像的系统功能。

【功能描述】

本项目将设计一款动态图形图像程序，通过直接运行来实现动画的效果。

- 使用线性布局设计界面；
- 实现旋转、缩放、改变透明度等功能；
- 实现欢迎界面逐帧动画。

【基本框架】

基本框架如图 7.1 所示，将框架图转换成的效果如图 7.2、图 7.3 所示。

图 7.1　动态图形图像架图

图 7.2　动态图形图像主界面效果图

图 7.3　动态图形图像欢迎界面效果图

技能点 1 Bitmap 和 BitmapFactory

Bitmap 代表一个位图，也是最重要的图像处理类之一。它可以获取图像文件信息，进行图像剪切、旋转、缩放等操作，还可以指定格式保存图像文件。Bitmap 参数设置说明如表 7.1 所示。

表 7.1 Bitmap 的参数设置说明表

参数名称	说明
public void recycle()	回收位图占用的内存空间，把位图标记为 Dead
public final boolean isRecycled()	判断位图内存是否已释放
public final int getWidth()	获取图片的宽度值
public final int getHeight()	获取图片的高度值
public final boolean isMutable()	图片是否可修改
public int getScaledWidth(Canvas canvas)	获取指定密度转换后的图像的宽度
public int getScaledHeight(Canvas canvas)	获取指定密度转换后的图像的高度
public boolean compress(CompressFormat format, int quality, OutputStream stream)	按指定的图片格式以及画质 format：图片格式 quality：图片质量 stream：将照片转换为输出流

Bitmap 提供了一些静态方法来创建新的 Bitmap，如表 7.2 所示。

表 7.2 Bitmap 静态类方法

静态方法	描述
createBitmap(Bitmap source,int x,int y,int width,int height)	创建位图 Source: 指定坐标点（x，y） width：宽，height：高
createScaledBitmap(Bitmap src,int dstWidth,int dstHeight,Boolean filter)	缩小位图比例 Src：源位图资源 dstWidth：缩放后的宽 dstHeight：缩放后的高 filter：filter 为 true 时过滤图片

BitmapFactory 是一个工具类，提供了大量的方法，用于解析来源不同的数据以创建

Bitmap 对象。BitmapFactory 类方法如表 7.3 所示。

表 7.3　BitmapFactory 类方法

BitmapFactory 类方法	描述
decodeByteArray(byte[] data,int offset, int length)	从数组中读取图片 Data：准备进行编译的资源数据 Offset：位移量，一般为 0 Lenght：编译数据的长度
decodeFile(String pathName)	从文件中读取图片 String pathName：准备解析的数据文件
decodeResource(Resource res,int id)	从资源中读取图片 Res：加载的位图资源文件的对象 Id：加载的位图资源的 Id
decodeStream(InputStream is)	将输入流转换为图片 Is：输入流中解析创建 Bitmap 对象

大多数情况只需将图片存放于 /res/drawable/ 目录下，便可在程序中通过图片所对应的 ID 进行封装使用。但由于手机系统内存较小，如果系统不断解析，创建 Bitmap 对象，将会导致程序运行时引发 OOM 错误。

拓展：想了解或学习 BitmapFactory 工具类的知识，可扫描下方二维码，获取更多信息。

技能点 2　逐帧动画

1　逐帧动画简介

逐帧动画是比较容易理解的动画，程序员需将多张连续的静态图片进行收集，然后由 Android 来控制这些图片显示的顺序和时间，利用肉眼“视觉暂留”的原理，实现“动画”的错觉。

2　逐帧动画格式及标签

（1）格式

逐帧动画 XML 文件创建位置如图 7.4 所示。

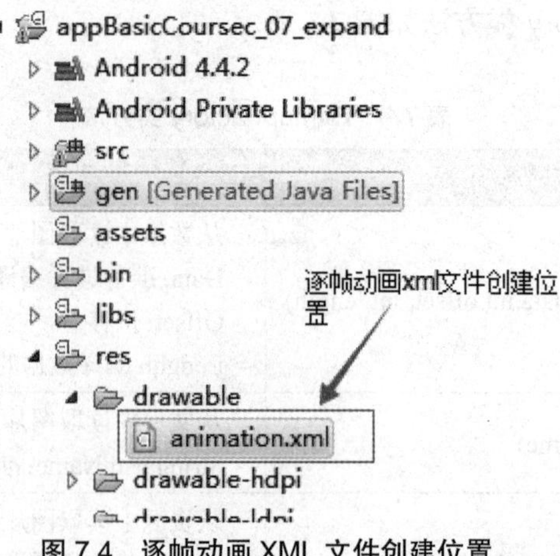

图 7.4　逐帧动画 XML 文件创建位置

定义逐帧动画很简单，在如图 7.4 所示位置创建 animation 文件，<animation-list…/> 元素中使用 <item…/> 子元素定义动画的全部帧，并指定各帧的持续时间即可。具体实现方法如下所示：

```
<?xml version="1.0" encoding="utf-8"?>
<animation-list xmlns="http://schemas.android.com/apk/res/android"
    android:oneshot=["true"|"false"] >
    <item android:drawable="@[package:]drawable/drawable_resource_name"
        android:duration="integer" />
</animation-list>
```

（2）标签

逐帧动画是一种常见的动画形式，原理是从“连续的关键帧”中分解动画动作，就是在时间轴的每帧上逐帧绘制不同的内容，让它连续播放从而形成动画。逐帧动画标签如表 7.4 所示。

表 7.4　帧动画标签

标签	属性值	说明
<animation-list>	android:oneshot: 为 true，该动画只播放一次，停止在最后一帧	包含若干 <item> 标记
<item>	android:drawable: 图片帧的引用 android:duration: 图片帧的停留时间 android:visible: 图片帧是否可见	每个 <item> 标记定义了一个图片帧，包含图片资源的引用等属性

3　AnimationDrawable 类的常用方法

实现图片逐帧播放的功能，要使用 Android 提供的 AnimationDrawable 类，该类作为某一个 View 的 background 来使用。在 AnimationDrawable 类中提供 addFrame() 函数为动画添加帧图片，但为了便于管理，建议使用 XML 文件来为该动画进行配置。AnimationDrawable 类的常用方法说明如表 7.5 所示。

表 7.5　AnimationDrawable 类的常用方法说明

类名	说明
mAnimationDrawable.addFrame(Drawable frame, int duration)	添加一个帧动画
mAnimationDrawable.getDuration(int i)	获得帧动画的时间
mAnimationDrawable.getFrame(int index)	获得指定索引的 Drawable 对象
mAnimationDrawable.getNumberOfFrames()	获得帧动画的总数
mAnimationDrawable.isOneShot()	帧动画是否只运行一次
mAnimationDrawable.isRunning()	帧动画是否处于运行状态
mAnimationDrawable.setOneShot(boolean onsShot)	设置帧动画是否只运行一次
mAnimationDrawable.setVisible(boolean visible, boolean restart)	设置帧动画是否可见
mAnimationDrawable.start()	运行帧动画
mAnimationDrawable.stop()	停止帧动画

4　逐帧动画实现

（1）编写逐帧动画 XML 文件。

```
<?xml version="1.0" encoding="utf-8"?>
<animation-list xmlns:android="http://schemas.android.com/apk/res/android"
 android:oneshot="true">
     <item android:drawable="@drawable/no1" android:duration="1000" />
     <item android:drawable="@drawable/no2" android:duration="1000" />
     <item android:drawable="@drawable/no3" android:duration="1000" />
     <item android:drawable="@drawable/no4" android:duration="1000" />
 </animation-list>
```

（2）使用 AnimationDrawable 类实现逐帧动画。

```
ImageView rocketImage = (ImageView) findViewById(R.id.iv);
        // 获取 iv 的背景
        AnimationDrawable rocketAnimation = (AnimationDrawable) rocketImage.get-
Background();
        // 开始播放
        rocketAnimation.start();
```

（3）实现效果如图 7.5 所示。

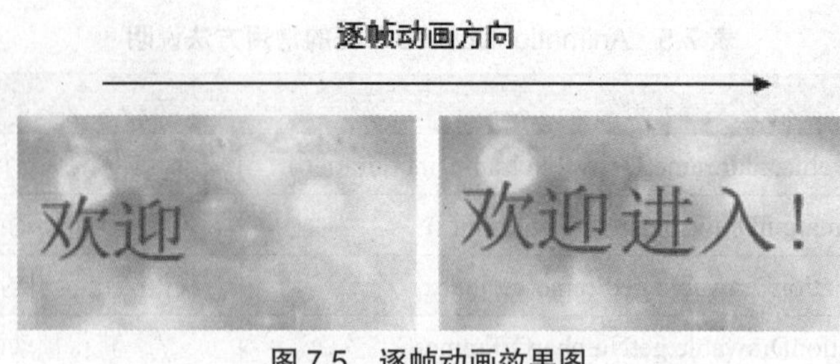

图 7.5 逐帧动画效果图

技能点 3 补间动画

1 补间动画简介

补间动画指程序员需指定动画的开始和结束等"关键帧"，而动画变化的"中间帧"，由系统计算并补齐，这也是作者将 Tween 动画翻译为"补间动画"的原因。图 7.6 为作补间动画的示意图。

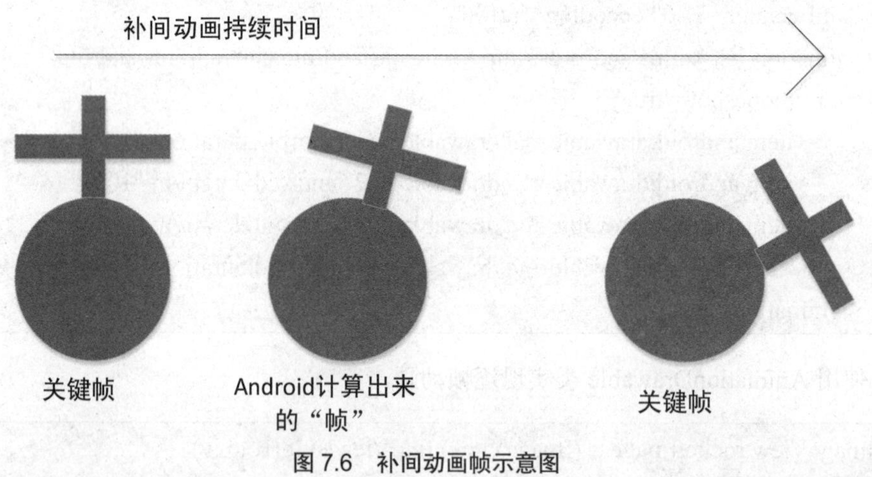

图 7.6 补间动画帧示意图

2 补间动画分类

补间动画分为两大类，分别是形状补间动画和动作补间动画。两个动画直接的区别如下：

- 形状补间动画是在 Flash 的时间帧面板上，在一个关键帧上绘制一个形状，更改该形状或绘制另一个形状将其设为另一个关键帧。Flash 将自动根据二者之间的帧的值或形状来创建动画，它可以实现多种变化，如两个图形之间颜色、形状、大小、位置的相互变化。
- 动作补间动画是指在 Flash 的时间帧面板上，在一个关键帧上放置一个元件，更改这个元件的大小、颜色、位置、透明度等将其设为另一个关键帧，Flash 将自动根据二者之间帧的值创建动画。

图 7.7 是缩放动作补间动画示意图，图 7.8 是位移补间动画示意图。

（a）执行缩放的补间动画前

（b）执行缩放的补间动画后

图 7.7　缩放的补间动画示意图

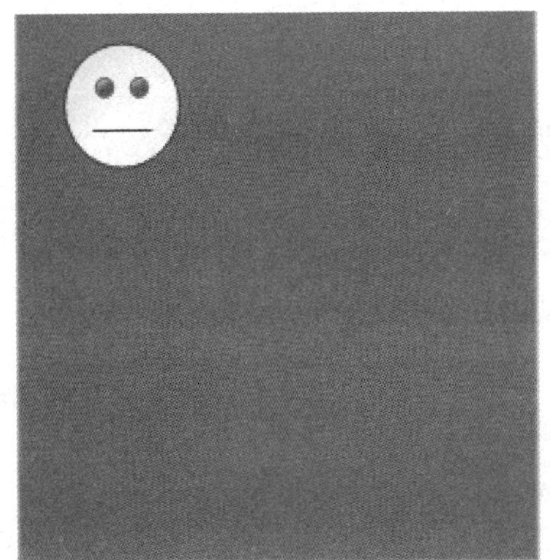

（a）执行位置改变的补间动画前

（b）执行位置改变的补间动画后

图 7.8　改变位置的补间动画示意图

3　补间动画插值器

在补间动画中，一般只定义关键帧（首帧或尾帧），中间帧是不需要自己设置的，系统会自动生成，生成中间帧的这个过程可以称为“插值”。插值器定义了动画变化的速率，提供值随时间变化而变化，比如加速、减速等。几种常见的插值器如表 7.6 所示。

表 7.6 补间动画常见的插值器

Interpolator 对象	资源 ID	功能作用
AccelerateDecelerateInterpolator	@android:anim/accelerate_decelerate_interpolator	先加速再减速
AccelerateInterpolator	@android:anim/accelerate_interpolator	加速
AnticipateInterpolator	@android:anim/anticipate_interpolator	先回退一小步，后加速前进
AnticipateOvershootInterpolator	@android:anim/anticipate_overshoot_interpolator	在上一基础超出终点一小步，再回到终点
BounceInterpolator	@android:anim/bounce_interpolator	最后阶段弹球效果
CycleInterpolator	@android:anim/cycle_interpolator	周期运动
DecelerateInterpolator	@android:anim/decelerate_interpolator	减速
LinearInterpolator	@android:anim/linear_interpolator	匀速
OvershootInterpolator	@android:anim/overshoot_interpolator	快速到达终点并超出一小步，最后回到终点

4 补间动画实现

（1）初始化 ImageView 并创建。

```
iv = (ImageView) findViewById(R.id.img_iv);// 初始化 ImageView
scale();// 创建缩放方法
alpha();// 创建透明度方法
```

（2）在缩放方法与透明度方法中设置补间动画各类参数并开启动画。

```
//  透明度
  public void alpha(){
  oa3 = ObjectAnimator.ofFloat(iv, "alpha", 1, 0.2f);
  oa3.setDuration(1600);// 动画时间
  oa3.setRepeatCount(1);// 动画次数
  oa3.setRepeatMode(ValueAnimator.REVERSE);// 动画方式
  oa3.start();                    // 动画开始
  }
//  缩放
public void scale(){
  oa2 = ObjectAnimator.ofFloat(iv, "scaleX", 2f, 0.7f, 3f, 0.7f);
  oa2.setDuration(1600);
  oa2.setRepeatCount(1);
```

```
oa2.setRepeatMode(ValueAnimator.REVERSE);
oa2.start();
}
```

（3）实现效果如图 7.9 所示。

图 7.9　补间动画效果图

技能点 4　属性动画

1　属性动画简介

属性动画是一种动画框架系统，能满足大部分动画需求。属性动画能在动画执行的过程中改变它的任意属性值，所以不会影响其在动画执行后所在位置的正常使用。

2　属性动画的优点及定义方式

（1）优点

- 补间动画只能定义两个关键帧的“透明度”“旋转”“缩放”“位移”4 个属性的变化，而属性动画可定义任何属性的变化。
- 补间动画只能对 UI 组件执行动画，而属性动画可对任何对象执行动画（不论是否显示在界面）。

（2）定义方式

- 使用 ValueAnimator() 或 ObjectAnimator() 的静态工厂方法创建动画。
- 使用文件资源定义动画。

3　属性动画和补间动画区别

补间动画与属性动画在视觉上的效果是相同，但实际有很大区别。属性动画优点很多，当属性动画移动后，如果不再回到起始的位置，那么点击执行动画后的新位置，将接收不到 Click 事件（点击事件）。补间动画只实现了图像位置的改变，但控件实际上并未发生位移，点击起始位置则可以接收到点击事件。

补间动画通过不断地调用 OnDraw() 方法来进行 UI 的绘制，而属性动画一般只调用 ViewGroup() 进行绘制。属性动画执行结束后不会主动恢复到原来的状态，它会一直保持最后的状态，直到下一次执行改变的时候才改变状态。为了增加动画的灵活性，属性动画通过 ObjectAnimator() 和 PropertyValueHolder() 进行动态控制。用一个例子具体讲解一下属性动画和补间动画的区别，如图 7.10、图 7.11 所示。

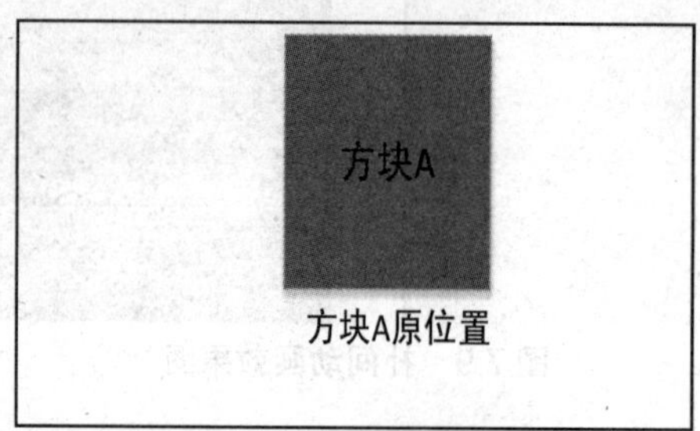

图 7.10　执行动画前位置

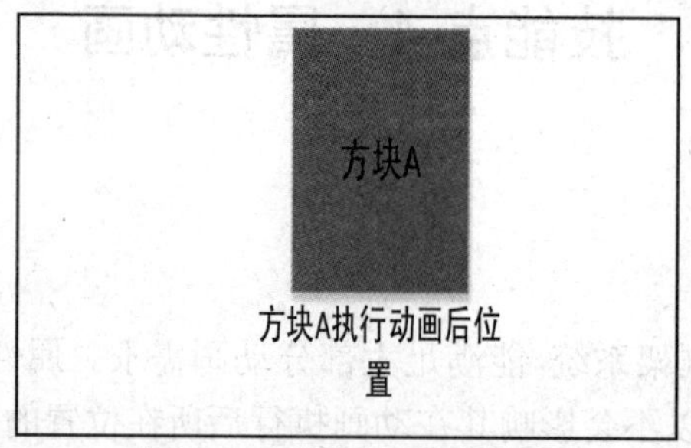

图 7.11　执行动画后位置

方块 A 是执行属性动画移动到之后的位置的话，该方块 A 就实际真的在移动后的位置了。但是如果方块 A 是执行补间动画移动到之后的位置的话，该方块 A 的实际位置还在原位置，只不过是视觉上觉得它的位置在执行动画后的位置而已。

4　属性动画实现

（1）初始化 ImageView 控件并创建旋转方法。

```
iv = (ImageView) findViewById(R.id.img_iv);// 初始化 ImageView
rotate();// 创建旋转方法
```

（2）在旋转方法中设置属性动画参数，并开始动画。

```
    // 旋转
        public void rotate(){
            ra = new RotateAnimation(0, -720, Animation.RELATIVE_TO_SELF, 0.5f,
Animation.RELATIVE_TO_SELF, 0.5f);  // 设置旋转起始结束角度以及旋转方向
            ra.setDuration(1600);// 动画时间
            ra.setRepeatCount(1);// 动画次数
            ra.setRepeatMode(Animation.REVERSE);// 动画方式
            ra.setFillAfter(true);
            iv.startAnimation(ra);// 动画开始
        }
```

（3）实现效果如图 7.12 所示。

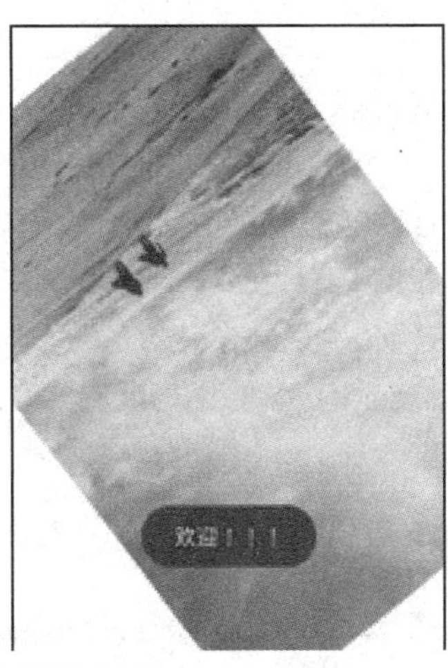

图 7.12　属性动画效果图

第一步：在 Eclipse 中创建一个 Android 工程，命名为“动态图形图像系统”，并设计界面。如图 7.2、图 7.3 所示。

第二步：在 src 文件夹下建立 MainActivity.java 文件，并实现界面初始化，创建补间动画以及属性动画方法。具体如代码 CORE0701 所示。

代码 CORE0701：界面初始化及补间动画，属性动画的创建

```
/*
 * 1 此处填写界面初始化代码并创建补间动画与属性动画方法
 */
        iv = (ImageView) findViewById(R.id.img_iv);
```

```
            Toast.makeText(getApplicationContext(), " 欢迎!!! ",
            Toast.LENGTH_LONG).show();
            rotate();// 属性动画旋转
            scale();// 补间动画缩放
            alpha();// 补间动画透明度
```

第三步:实现补间动画以及属性动画。具体如代码 CORE0702 所示。

代码 CORE0702:补间动画以及属性动画

```
/*
    * 2 此处填写补间动画
    * 透明度与缩放动画
    */
//  透明度
    public void alpha(){
        oa3 = ObjectAnimator.ofFloat(iv, "alpha", 1, 0.2f);
        oa3.setDuration(1600);// 动画时间
        oa3.setRepeatCount(1);// 动画次数
        oa3.setRepeatMode(ValueAnimator.REVERSE);// 动画方式
        oa3.start();// 动画开始
    }
//  缩放
    public void scale(){
        oa2 = ObjectAnimator.ofFloat(iv, "scaleX", 2f, 0.7f, 3f, 0.7f);
        oa2.setDuration(1600);
        oa2.setRepeatCount(1);
        oa2.setRepeatMode(ValueAnimator.REVERSE);
        oa2.start();
    }

    /*
    * 3 此处填写属性动画
    * 旋转动画
    */
//  旋转
public void rotate(){
            ra = new RotateAnimation(0, -720, Animation.RELATIVE_TO_SELF,
0.5f, Animation.RELATIVE_TO_SELF, 0.5f);        // 设置旋转起始结束角度以及旋转方向
            ra.setDuration(1600);// 动画时间
```

```
            ra.setRepeatCount(1);// 动画次数
            ra.setRepeatMode(Animation.REVERSE);// 动画方式
            ra.setFillAfter(true);
            iv.startAnimation(ra);// 动画开始
    }
```

第四步：实现延时跳转功能。具体如代码 CORE0703 所示。

代码 CORE0703：延时跳转

```
/*
     * 4 此处填写延时跳转代码
     */
    Thread thread = new Thread() {
        public void run() {
            try {
                long ms = 0;// 定义时间长度
                while (m_bSplashActive && ms < m_dwSplashTime) {
                    sleep(30);

                    if (!m_bPaused)
                        ms += 30;
                    // 判断当前状态以及时间长度
                }
                startActivity(new android.content.Intent("MainActivity"));
            } catch (Exception ex) {
                Log.e("Splash", ex.getMessage());
            } finally {
                Intent intent = new Intent(MainActivity.this,
                        MenuActivity.class);
                startActivity(intent);
                finish();
                // 进行跳转
            }
        }
    };
// 启动线程
    thread.start();
}
```

第五步：实现欢迎界面逐帧动画。具体如代码 CORE0704 所示。

代码 CORE0704：欢迎界面动画

```
ImageView rocketImage = (ImageView) findViewById(R.id.iv);
            // 获取 iv 的背景
            AnimationDrawable rocketAnimation = (AnimationDrawable)
rocketImage.getBackground();
            // 开始播放
            rocketAnimation.start();
```

第六步：运行程序，结果如图 7.2、图 7.3 所示。

【拓展目的】

熟练掌握各类动画的使用方法与技能。

【拓展内容】

实现手机扫描杀毒功能。效果如图 7.13 所示。

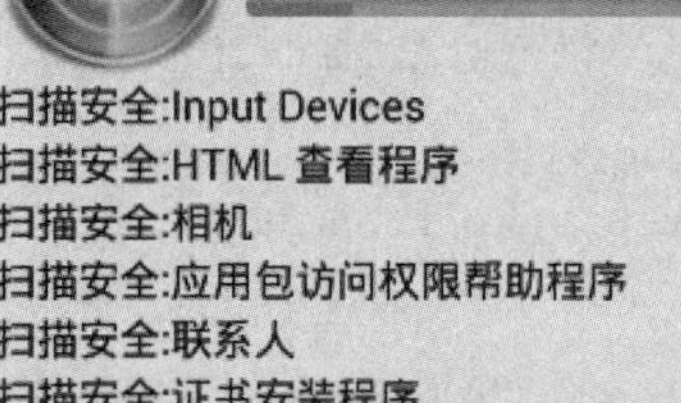

图 7.13 手机扫描杀毒系统主界面效果图

【拓展步骤】

（1）设计思路：添加旋转动画功能，实现杀毒扫描动画以及进度条动画。

（2）实现界面初始化。具体如代码 CORE0705 所示。

代码 CORE0705：界面初始化

```
/*
 * 1 此处填写界面初始化代码
 */
        ivScanning = (ImageView) findViewById(R.id.iv_scanning);
        tvStatus = (TextView) findViewById(R.id.tv_status);
        pbProgress = (ProgressBar) findViewById(R.id.pb_progress);
        llContainer = (LinearLayout) findViewById(R.id.ll_container);
```

（3）实现扫描动画功能。具体如代码 CORE0706 所示。

代码 CORE0706：扫描动画

```
/*
 * 2 此处填写属性动画（旋转）代码
 */
        RotateAnimation anim = new RotateAnimation(0, 360,
                Animation.RELATIVE_TO_SELF, 0.5f,
Animation.RELATIVE_TO_SELF,
                0.5f);// 设置旋转角度，周期方向等参数
        anim.setDuration(2000);// 动画时间
        anim.setInterpolator(new LinearInterpolator());
        anim.setRepeatCount(Animation.INFINITE); // 设置动画重复次数
        ivScanning.startAnimation(anim);
```

（4）扫描手机文件，进行病毒查杀，并且实时更新进度条。具体如代码 CORE0707 所示。

代码 CORE0707：病毒查杀及进度条更新

```
/*
 * 3 此处填写病毒查杀代码
 */
        new Thread() {
            public void run() {
                SystemClock.sleep(2000);
                PackageManager pm = getPackageManager();
```

```
                List<PackageInfo> installedPackages = pm
        .getInstalledPackages(PackageManager.GET_UNINSTALLED_PACKAGES);
                pbProgress.setMax(installedPackages.size());// 设置进度条最大值
                int progress = 0;
                Random random = new Random();
// 4 判断查询的文件是否为病毒
                for (PackageInfo packageInfo : installedPackages) {
                    ScanInfo info = new ScanInfo();
                    String packageName = packageInfo.packageName;
                    String name = packageInfo.applicationInfo.loadLabel(pm)
                            .toString();
                    info.packageName = packageName;
                    info.name = name;
                    String apkPath = packageInfo.applicationInfo.sourceDir;
                    String md5 = MD5Utils.encodeFile(apkPath);

                    if (AntiVirusDao.isVirus(md5)) {
                        System.out.println(" 发现病毒 !");
                        info.isVirus = true;
// 更改 isVirus 信息
                        mVirusList.add(info);
                    } else {
                        System.out.println(" 扫描安全 !");
                        info.isVirus = false;
                    }
// 实时更新进度条数值
                    progress++;
                    pbProgress.setProgress(progress);
                    Message msg = Message.obtain();
                    msg.what = STATE_UPDATE_STATUS;
                    msg.obj = info;
                    mHandler.sendMessage(msg);

                    SystemClock.sleep(50 + random.nextInt(50)); // 延时操作
                }
                mHandler.sendEmptyMessage(STATE_SCAN_FINISH);
            };
        }.start();
```

```
        }
    private Handler mHandler = new Handler() {
            public void handleMessage(android.os.Message msg) {
                switch (msg.what) {
                // 更新界面
                case STATE_UPDATE_STATUS:
                    ScanInfo info = (ScanInfo) msg.obj;
                    tvStatus.setText(" 正在扫描 :" + info.name);
                    TextView view = new TextView(getApplicationContext());
                    if (info.isVirus) {
                        view.setText(" 发现病毒 :" + info.name);
                        view.setTextColor(Color.RED);
                    } else {
                        view.setText(" 扫描安全 :" + info.name);
                        view.setTextColor(Color.BLACK);
                    }
                    llContainer.addView(view, 0);
                    break;
                case STATE_SCAN_FINISH:
                    tvStatus.setText(" 扫描完毕 ");
                    ivScanning.clearAnimation();

                    if (!mVirusList.isEmpty()) {

                        // 有病毒进行提示
                        showAlertDialog();
                    } else {
                        Toast.makeText(getApplicationContext(), " 您的手机很安全 ,
请放心使用 !", 0).show();
                    }
                    break;

                default:
                    break;
                }
            };
        };
    // 病毒判断
```

```
public class AntiVirusDao {
    private static final String PATH =
"/data/data/com.example.appbasiccoursec_07_expand/cache/antivirus.db";

    public static boolean isVirus(String md5) {
        SQLiteDatabase database = SQLiteDatabase.openDatabase(PATH, null,
                SQLiteDatabase.OPEN_READONLY); // 打开数据库

        Cursor cursor = database.rawQuery("select * from datable where md5=?",
                new String[] { md5 });// 将查询的信息添加到游标中

        boolean isVirus = false;
        if (cursor.moveToFirst()) {
            isVirus = true;
        }
        cursor.close();
        database.close();
        return isVirus;
    }
}
```

(5) 运行程序，结果如图 7.13 所示。

本项目主要介绍了 Android 的图形图像处理。Android 提供了逐帧动画、补间动画、属性动画支持，需要重点掌握。可以根据程序需要，引用或绘制各种各样的图形，丰富界面的多样性。

start	开始
head	头文件
body	文件体
create	建立
final	终结的
fill	填充

listener　　收听者
class　　类
button　　按键

一、选择题

1. WVGA 所指的分辨率为（　　）。

A. 480×800　B. 480×854　C. 320×480　D. 240×320

2. 大分辨率的图片应存放在（　　）文件夹中。

A. drawable-hdpi　B. drawable-mdpi　C. drawable-ldpi　D. drawable

3. 属性动画中 android:duration 指的是（　　）。

A. 动画重复次数　B. 动画差值方式　C. 动画持续时间　D. 重复行为

4. R.drawable.filename 是一个（　　）类型的常量。

A.string　B.float　C. int　D. bitmap

5. 下列 Bitmap 参数获取图片宽度值的是（　　）。

A.. int outWidth　B. int inDensity　C. boolean inScaled　D. int inSampleSize

二、填空题

1. Android 类 Interpolator 中设置 ________ 可以使动画以均匀的速度变化。

2. 补间动画中 AlphaAnimation 设置透明度的范围由 _____ 到 _____ 。

3. AnimationDrawable 代表的动画使用 _______ 开始，使用 ______ 结束。

4. Android 系统提供了 ______ 显示普通静态图片。

5. Android 应用中的图片不仅包括 *.png、*.jpg、*.gif 等各种格式的位图，也包括 _________ 定义的各种 Drawable 资源。

三、判断题

1. 使用 XML 文件来为该动画配置图片，这样更简单而且易于管理。（　　）

2. 只能使用 ObjectAnimator 的静态工厂方法创建动画。（　　）

3. 属性动画执行后不会影响其在动画执行后所在位置的正常使用。（　　）

4. 补间动画只能对 UI 组件执行动画，而属性动画可对任何对象执行动画。（　　）

5. mAnimationDrawable.setVisible(boolean visible, boolean restart) 可以设置图片的透明度。（　　）

四、简答题

1. Bitmap 和 BitmaDrawable 之间应如何转换？试说明。

2. drawable- hdpi、drawable- mdpi、drawable-ldpi 有什么区别？

五、上机题

构造一个 ImageView，从 drawable 资源中调用并添加到布局中。

```
protected void onCreate(Bundle savedInstanceState) {
super.onCreate(savedInstanceState);
// Create a LinearLayout in which to add the ImageView mLinearLayout = new Linear-
Layout(this);
// Instantiate an ImageView and define its properties
ImageView i = new ImageView(this);
i.setImageResource(R.drawable.my_image);
i.setAdjustViewBounds(true); // set the ImageView bounds to match the Drawable's di-
mensions
i.setLayoutParams(new
Gallery.LayoutParams(LayoutParams.WRAP_CONTENT,          LayoutParams.WRAP_
CONTENT));
// Add the ImageView to the layout and set the layout as the content view
mLinearLayout.addView(i);
setContentView(mLinearLayout); }
```

示例 XML

下面的 XML 片断显示了如何添加一个可绘制资源到一个 XML 布局中的 ImageView 里。

```
<ImageView
android:layout_width="wrap_content"
android:layout_height="wrap_content"
android:tint="#55ff00"
android:src="@drawable/my_image"/>
```

项目八　Service 服务

通过实现音乐的播放、停止、暂停、重播等功能，学习 Android 四大组件中 Service 生命周期相关知识，掌握 Service 启动、绑定、退出等使用方法。在任务实现过程中：

- 了解 Service 的特点；
- 掌握 Service 的启动方式；
- 掌握生命周期使用方法；
- 掌握 Service 的两种类型：本地服务和远程服务。

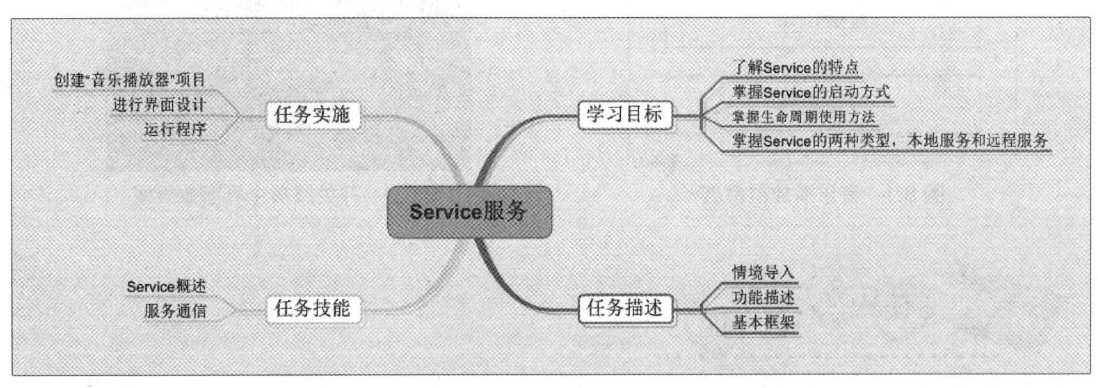

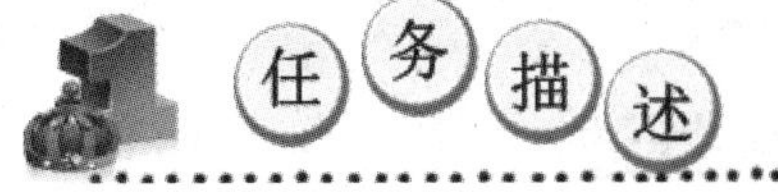

【情景导入】

日常生活中通过手机播放音频来学习、工作和娱乐已经成为大多数人采取的主要途径，本项目以手机音乐播放器为业务背景，通过 Service 组件技术，实现播放器运行过程中的核心功能：播放、暂停、重播、停止等。

【功能描述】

本项目将设计一款音乐播放器程序。

- 使用线性布局设计音乐播放器主界面；
- 初始化进度条的长度，获取音乐文件的长度；
- 点击“播放”按钮，播放音乐并且更新进度条；
- 点击“暂停”按钮，暂停播放音乐和更新进度条；
- 点击“重播”按钮，音乐重新开始播放并更新进度条；
- 点击“停止”按钮，停止播放音乐和更新进度条。

【基本框架】

基本框架如图 8.1 所示，将框架图转换成的效果如图 8.2 所示。

图 8.1 音乐播放器框架

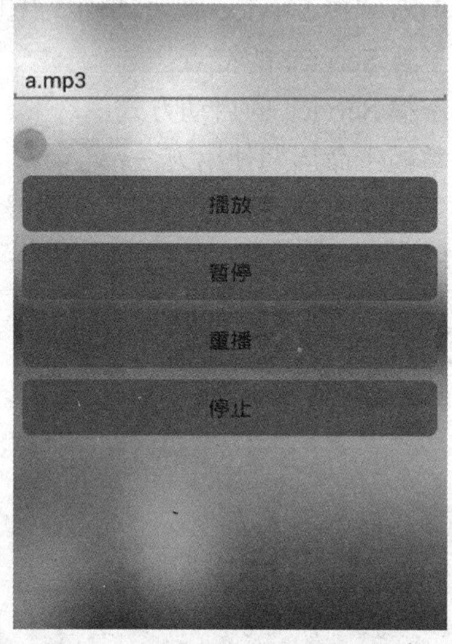

图 8.2 开关服务主界面效果图

技能点 1 Service 概述

1 Service 简介

Service 是可以在后台执行长时间操作而不使用用户界面的应用组件，与 Android 四大组件中的 Activity 最相似，代表着可执行程序。Service 有自己的生命周期，按运行类别分类可分为前台 Service 与后台 Service 两种。Service 主要有两个功能：后台持续运行，跨进程访问。Service 与 Activity 的不同点在于：Service 一直都在后台运行，没有用户界面。一旦 Service 被启动，便具有自己的生命周期。如果某个程序组件不需要在运行时与用户交互或者显示界面，则使用 Service。当系统出现内存不足情况时就有可能会回收掉正在后台运行的 Service，因为后台 Service 的系统优先级比较低。用前台 Service 可以让 Service 一直保持运行状态，防止由于系统内存不足的原因导致被回收。

前台 Service 和后台 Service 最大的区别在于前台 Service 会在系统的状态栏显示一个一直运行的图标。当下拉状态栏后会看到更为详细的信息，类似于通知。有些项目要实现 Service 在后台更新数据的同时，还要在状态栏显示图标和新的信息的时候必须使用前台 Service 了。前台 Service 应用如图 8.3 所示。

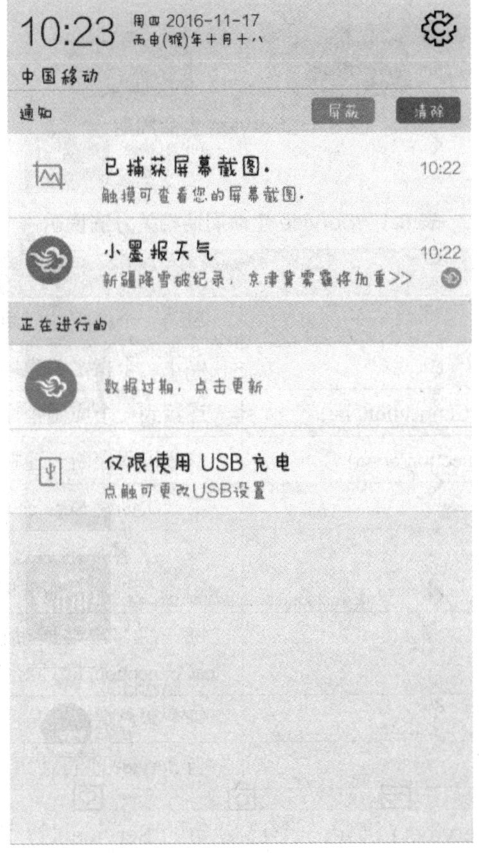

图 8.3　前台 Service 应用

2　Service 方法说明

Service 有自己的生命周期，经历了创建到销毁的过程，Service 有两种启动方式：startService() 和 bindService()。Service 生命周期如图 8.4 所示。Service 生命周期相关方法说明如表 8.1 所示。

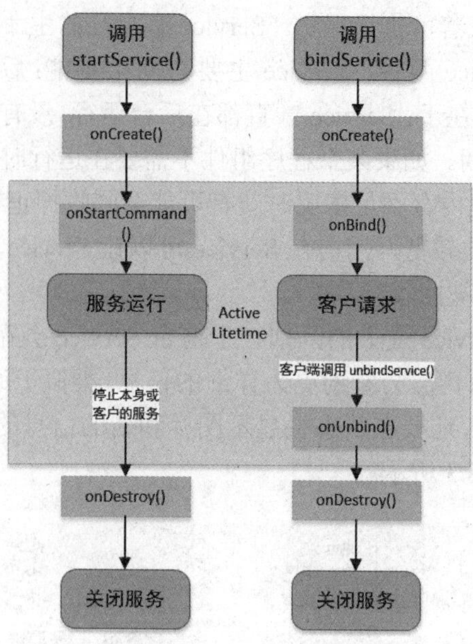

图 8.4 Service 生命周期

表 8.1 Service 生命周期相关方法说明

名称	说明
startService(Intent service)	启动一个指定的应用程序服务
stopService(Intent service)	停止一个指定的应用程序服务
bindService(Intent, ServiceConnection, int)	连接到一个应用程序服务
unbindService(ServiceConnection conn)	从应用程序断开连接服务
onCreate()	第一次创建 Service 时执行该方法
onStartCommand()	每一次客户端通过调用 startService(Intent service) 显示地启动服务时执行该方法
onBind()	每一次客户端通过调用 bindService(Intent, ServiceConnection, int) 隐形地启动服务时执行该方法
onUnbind()	每个客户端断开与服务的绑定时执行该方法
onDestory()	当 Service 不再使用，并已被删除时执行该方法

当 Activity 调用 bindService() 绑定一个已启动的 Service 时，Sercive 的生命周期不会和 Activity 相同的原因是系统把 Service 内部 IBinder 对象传给 Activity，只要 Activity 调用 unBindService() 方法就可以取消与该 Service() 的绑定时，就切断该 Activity 与 Service 之间的关联。但是注意一点，此时并不能停止 Service 组件运行。

Service 的生命周期要比 Activity 的生命周期简单得多，只继承了三个方法：onCreate()、onStart()、onDestroy()。第一次使用 startService() 方法（如图 8.5 所示）启动 Service 时，会依次调用 onCreate()、onStart() 这两个方法，当停止 Service 时，就会执行 onDestroy() 方法。这里需要

注意的是，如果当前 Service 已经启动了，当我们再次启动 Service 时，不会再执行 onCreate() 方法，而是直接执行 onStart() 方法。

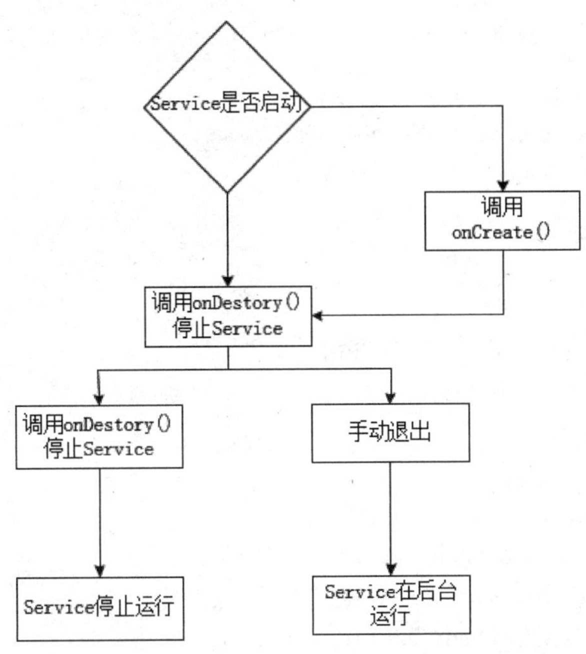

图 8.5　调用 startService() 方法启动 Service

注意：Service 是运行在所在进程的主线程中，Service 不会创建自己的 thread，也不会运行在单独的进程中（除非另外指定）。因此，如果做 MP3 播放或视频播放这些消耗 CPU 的工作或阻塞型的操作，就一定要在 Service 中创建一个新的线程，这个线程就来做那些工作。这样就会减少应用没有反应（ANR）错误。Activity 对用户操作的快速反应也是通过开启新线程做耗内存操作来实现。

3　Service 实现

虽然 Service 在后台运行，但是 Service 后端的数据还是要用户可见的，最终还是要呈现在前端 Activity 上的，因此在启动 Service 时，系统会重新开启一个新的进程。当想获取启动的 Service 实例时，可以用到 bindService() 和 onBindService() 方法。在这两种方法中分别执行了 Service 中的 IBinder() 和 onUnbind() 方法。接下来新建一个 Service，命名为 Service.java，具体实现方法如下所示：

```
package com.tutor.servicedemo;
public class Service extends Service {
    // 定义一个 Tag 标签
    private static final String TAG = "MyService";
    // 这里定义一个 Binder 类，用在 onBind() 方法里，这样 Activity 那边可以获取到
```

```
private MyBinder mBinder = new MyBinder();
// 绑定
@Override
public IBinder onBind(Intent intent) {
        Log.e(TAG, "start IBinder~~~");
        return mBinder;
}
@Override
public void onCreate() {
        Log.e(TAG, "start onCreate~~~");
        super.onCreate();
}
// 开启服务
@Override
public void onStart(Intent intent, int startId) {
        Log.e(TAG, "start onStart~~~");
        super.onStart(intent, startId);
}
// 关闭服务
@Override
public void onDestroy() {
        Log.e(TAG, "start onDestroy~~~");
        super.onDestroy();
}
// 取消绑定
@Override
public boolean onUnbind(Intent intent) {
        Log.e(TAG, "start onUnbind~~~");
        return super.onUnbind(intent);
}
// 这里是一个获取当前时间的函数
public String getSystemTime(){
        Time t = new Time();
        t.setToNow();
        return t.toString();
}
public class MyBinder extends Binder{
```

```
                MyService getService() {
                        return MyService.this;
                }
        }
}
```

拓展：想了解可学习 Android Service 的生命周期的详情，可扫下方二维码，获取更多信息。

技能点 2　服务通信

1　本地服务通信

本地服务通信是最常用的后台 Service，用于实现应用程序内部的一些耗时任务，比如查询升级信息，并不占用应用程序比如 Activity 所属线程，而是单开线程后台执行。

Service 与访问者之间无法进行通信以及数据交换，若 Service 和访问者之间要进行通信，就调用 bindService() 和 unBindService() 这两个方法来启动或关闭 Service。

Context 的 bindService() 方法的完整方法为 bindService(Intent service, ServiceConnection conn, int flags)。Context 的 bindService() 方法参数说明如表 8.2 所示。

表 8.2　Context 的 bindService() 方法参数说明

名称	说明
service	通过 Intent 指定要启动的 Service
conn	该参数是一个 ServiceConnection 对象，该对象用于监听访问者与 Service 之间的连接情况
flags	指定绑定时是否自动创建 Service

注意：当调用者主动通过 unBindService() 方法断开与 Service 的连接时，Service Connection 对象的 onServiceDisconnected(ComponentName name) 方法并不会被调用。

使用 startService() 启动服务后，要使用 stopService() 停止 Service。同时使用 startService() 与 bindService() 要注意，需要 unbindService() 与 stopService() 同时调用，才能终止 Service。

2 远程服务通信

远程服务是一个独立的进程，它不受其他进程的影响，能为其他应用程序提供可用的接口——进程间通信 IPC（Inter-Process Communication），Android 提供了 AIDL（Android Interface Definition Language，接口描述语言）工具来帮助进程间接口的建立。

远程服务通信适用于为其他应用程序提供公共服务的 Service，这种 Service 就是系统常驻 Service。

当创建远程 Service 时，首先要通过 AIDL 文件定义 Service 向客户端（Client）提供的接口，在对应的目录下添加一个后缀为 .aidl 的文件，IMyAidlInterface.aidl 文件内容如下所示：

```
//IMyAidlInterface.aidl 文件内容
package com.zihao.remoteservice.server;interface IMyAidlInterface {
    String getMessage();
}
```

Aidl 的适用场景为：只有允许客户端从不同的应用程序去访问 Service 时，可以使用 Aidl 来实现。

当我们创建远程 Service 时，还需要新建 Remote Service，在远程服务中，通过 Service 的 onBind()，在客户端与服务建立连接时，用来传递 Stub（存根）对象。具体代码如下所示：

```
// 远程服务示例
public class RemoteService extends Service {
    public RemoteService() {
    }
    @Override
    public IBinder onBind(Intent intent) {
        return stub;
    // 在客户端连接服务端时，Stub 通过 ServiceConnection 传递到客户端
    }
    // 实现接口中暴露给客户端的 Stub--Stub 继承自 Binder，它实现了 IBinder 接口
    private IMyAidlInterface.Stub stub = new IMyAidlInterface.Stub(){
    // 实现了 AIDL 文件中定义的方法
        @Override
        public String getMessage() throws RemoteException {
    // 在这里我们只是用来模拟调用效果，因此随便反馈值给客户端
            return "Remote Service 方法调用成功 ";
        }
    };
}
```

同时，在 AndroidManifest.xml 中对 Remote Service 进行如下配置：

```
<service
    android:name=".RemoteService"
    android:process="com.test.remote.msg">
    <intent-filter>
        <action android:name="com.zihao.remoteservice.RemoteService"/>
    </intent-filter>
</service>
```

第一步：在 Eclipse 中创建一个 Android 工程，命名为“音乐播放器”，并设计界面。如图 8.2 所示。

第二步：在 src 文件夹中建立 MainActivity.java 文件并实现界面初始化。

第三步：在 MainActivity 中实现获取音乐文件功能。具体如代码 CORE0801 所示。

代码 CORE0801：获取音乐文件以及点击事件

```
/**
    * 1 此处添加获取音乐文件代码
        */
     public void onClick(View v) {
            String pathway = path.getText().toString().trim();
            System.out.println(pathway+"------///////");
            File SDpath = Environment.getExternalStorageDirectory();
            File file = new File(SDpath, pathway);
            String path = file.getAbsolutePath();
            System.out.println(path + "1-----------");
            switch (v.getId()) {
            case R.id.bt_play:
                // 判断音乐文件的大小
                if (file.exists() && file.length() > 0) {
                mservice.play(path);            // 播放音乐
                    initSeekBar(); // 获取音乐长度
                    UpdateProgress();// 更新进度条进度
} else {
                    Toast.makeText(this, " 找不到音乐文件 ", 0).show();
```

```
                }
                break;
            case R.id.bt_pause:
//判断音乐文件长度
                if (file.exists() && file.length() > 0) {
                    mservice.pause();
                }else {
                    Toast.makeText(this, " 找不到音乐文件 ", 0).show();
                }
                break;
            case R.id.bt_replay:
//判断音乐文件长度
                if (file.exists() && file.length() > 0) {
                    mservice.replay(pathway);
                }else {
                    Toast.makeText(this, " 找不到音乐文件 ", 0).show();
                }
                break;
            case R.id.btn_stop:
                //停止音乐之前首先要退出子线程
                mThread.interrupt();
                if (mThread.isInterrupted()) {
                    mservice.stop();
                }
                break;
            default:
                break;
            }
    }
```

第四步:实现音乐播放功能。具体如代码 CORE0802 所示。

代码 CORE0802:音乐播放

```
//播放音乐
    @SuppressLint("NewApi")
    public void play(String path) {
        try {
            if (mediaPlayer == null) {
```

```
                Log.i(TAG, " 开始播放音乐 ");
                // 创建一个 MediaPlayer 播放器
                mediaPlayer = new MediaPlayer();
                mediaPlayer.create(this, R.raw.a);
                // 指定参数为音频文件
                mediaPlayer.setAudioStreamType(AudioManager.STREAM_MUSIC);
                // 指定播放的路径
                mediaPlayer.setDataSource(path);
                System.out.println(path+"==================");
                // 准备播放
                mediaPlayer.prepare();
                mediaPlayer.setOnPreparedListener(new OnPreparedListener() {
                    @Override
                    public void onPrepared(MediaPlayer mp) {
                        // TODO Auto-generated method stub
                        // 开始播放
                        mediaPlayer.start();
                    }
                });
            } else {
                int position = getCurrentProgress();
                mediaPlayer.seekTo(position);
                try {
                    mediaPlayer.prepare();
                } catch (Exception e) {
                    e.printStackTrace();
                }
                mediaPlayer.start();
            }
        } catch (Exception e) {
            e.printStackTrace();
        }
    }
```

第五步：实现音乐暂停功能。具体如代码 CORE0803 所示。

代码 CORE0803：音乐暂停

```
// 暂停音乐
```

```
        public void pause() {
            if (mediaPlayer != null && mediaPlayer.isPlaying()) {
                Log.i(TAG, " 播放暂停 ");
                mediaPlayer.pause(); // 暂停播放
            } else if (mediaPlayer != null && (!mediaPlayer.isPlaying())) {
                mediaPlayer.start();
            }
        }
```

第六步：实现音乐重播功能。具体如代码 CORE0804 所示。

代码 CORE0804：音乐重播

```
// 重新播放音乐
    public void replay(String path) {
        if (mediaPlayer != null) {
            Log.i(TAG, " 重新开始播放 ");
            mediaPlayer.seekTo(0);
            try {
                mediaPlayer.prepare();
            } catch (IllegalStateException e) {
                e.printStackTrace();
            } catch (IOException e) {

                e.printStackTrace();
            }
            mediaPlayer.start();
        }
    }
```

第七步：实现音乐停止功能。具体如代码 CORE0805 所示。

代码 CORE0805：音乐停止

```
// 停止音乐
    public void stop() {
        if (mediaPlayer != null) {
            Log.i(TAG, " 停止播放 ");
            mediaPlayer.stop();
            mediaPlayer.release();
            mediaPlayer = null;
```

```
        } else {
            Toast.makeText(getApplicationContext(), " 已停止 ", 0).show();
        }
    }
```

第八步：编写更新进度条的代码。具体如代码 CORE0806 所示。

代码 CORE0806：更新进度

```
/**
    * 3 此处添加更新进度条功能
        */
    private void initSeekBar() {
            // TODO Auto-generated method stub
    // 获取音乐长度添加到 mSeekBar 中
            int musicWidth = mservice.getMusicLength();
            mSeekBar.setMax(musicWidth);
        }
    private void UpdateProgress() {
            mThread = new Thread() {
                public void run() {
                    while (!interrupted()) {
                        // 调用服务中的获取当前播放进度
                        int currentPosition = mservice.getCurrentProgress();
                        System.out.println(currentPosition+"----currentPositi
on------");

                        Message message = Message.obtain();
                        message.obj = currentPosition;
                        message.what = 100;
                        handler.sendMessage(message);
                        }
                    };
                };
                mThread.start();
            }
            private Handler handler = new Handler() {
                public void handleMessage(android.os.Message msg) {
                    switch (msg.what) {
                    case 100:
```

```
        // 实时更新进度条信息
                                int currentPosition = (Integer) msg.obj;
                                mSeekBar.setProgress(currentPosition);
                                break;
                            default:

                            break;
                        }
                    };
                };
        // 获取资源文件的长度
        public int getMusicLength() {
            if (mediaPlayer != null) {
                return mediaPlayer.getDuration();
            }
            return 0;
        }

        // 获取当前进度
        public int getCurrentProgress() {
            System.out.println(mediaPlayer+"mediaPlayer");
            if (mediaPlayer != null & mediaPlayer.isPlaying()) {
                Log.i(TAG, " 获取当前进度 ");
                return mediaPlayer.getCurrentPosition();
            } else if (mediaPlayer != null & (!mediaPlayer.isPlaying())) {
                return mediaPlayer.getCurrentPosition();
            }
            return 0;
            }
```

第九步：编写取消绑定防止退出程序抱死的代码。具体如代码 CORE0807 所示。

代码 CORE0807：退出防抱死

```
    /**
        * 4 此处添加退出取消绑定功能防止抱死
            */
        private class myConn implements ServiceConnection {
                public   void   onServiceConnected(ComponentName   name,   IBinder
service) {
```

```
        // 获取绑定
        binder = (MyBinder) service;
                    mservice=binder.getService();
                }
                public void onServiceDisconnected(ComponentName name) {
                    conn=null;
                }
            }
        protected void onDestroy() {
            // 如果线程没有退出，则退出
            if (mThread != null & !mThread.isInterrupted()) {
                    mThread.interrupt();
                }
                unbindService(conn);
                super.onDestroy();
            }
```

第十步：运行程序。结果如图 8.2 所示。

【拓展目的】

熟悉并掌握 Service 生命周期的使用方法以及本地与远程服务。

【拓展内容】

实现"音乐播放器"上一曲下一曲功能。效果如图 8.6 所示。

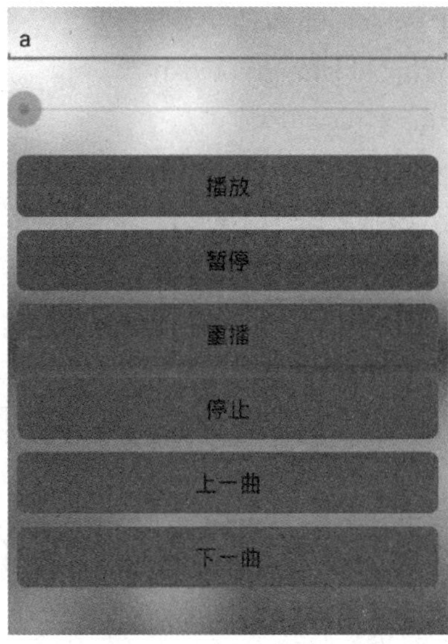

图 8.6　音乐播放器拓展主界面效果图

【拓展步骤】

(1)设计思路:添加"上一曲","下一曲"成功实现歌曲的切换功能。

(2)"上一曲","下一曲"功能。具体如代码 CORE0808 所示。

代码 CORE0808:"上一曲""下一曲"

```
/**
     * 1 此处添加切换单曲功能
         */
     public void onClick(View v) {
     switch (v.getId()) {
     case R.id.bt_before:
                 // 停止音乐之前首先要退出子线程
                 previous();
                 break;
             case R.id.btn_after:
                 // 停止音乐之前首先要退出子线程
                 next();
                 break;
         default:
                 break;
             }
         }
// 上一首
         private void previous() {
             if ((currIndex - 1) >= 0) {
                 mThread.interrupt();
                 if (mThread.isInterrupted()) {
                     mservice.stop();
                 }
// 设置播放音乐的名称,更新进度条进度
                 path.setText("a");
                 currIndex--;
                 System.out.println(currIndex + "-------currIndex-------");
                 mservice.play(name, currIndex);
                 UpdateProgress();
                 initSeekBar();
             } else {
```

```
                Toast.makeText(this, " 当前已经是第一首歌曲了 ",
    Toast.LENGTH_SHORT).show();
            }

        }
        // 下一自首
        private void next() {

            if (currIndex + 1  < 2) {
                mThread.interrupt();
                if (mThread.isInterrupted()) {
                    mservice.stop();
                }
    // 设置播放音乐的名称，更新进度条进度

                currIndex++;
                path.setText("b");
                System.out.println(currIndex + "-------currIndex-------");
                mservice.play(name, currIndex);
                UpdateProgress();
                initSeekBar();
            } else {
                Toast.makeText(this, " 当前已经是最后一首歌曲了 ", Toast.LEN
GTH_SHORT).show();
    //              Toast.makeText(MainActivity.this," 文 件 大 小 "+name.size(),  0).
show();
            }
        }

        // 监听器，当当前歌曲播放完时触发，播放下一首
        public void onCompletion(MediaPlayer mp) {
            if (name.size() > 0) {
                next();
            } else {
                Toast.makeText(this, " 播放列表为空 ", Toast.LENGTH_SHORT).
show();
            }
        }
```

学习 Service 需要重点掌握创建、配置 Service 组件，以及如何启动、停止 Service。本项目重点是 Service 的开发和通信，这个知识点需要重点掌握。另外，通过本项目的学习，需掌握 Android 的后台服务机制。

service	服务
stop	停止
define	定义
error	错误
float	单精度浮点
index	索引
thread	线程
runable	可捕获的
title	标题
try	试图

一、选择题

1. Service 的生命周期不包含（　　）。

A. onCreate()　　B. onStart()　　C. onDestroy()　　D. onStop()

2. 调用 bindService 的生命周期为：onCreate() → onBind() → onUnbind() → onDestory()。（　　）有且只能调用一次。

A. onCreate()　　B. onUnbind()　　C. onBind()　　D. onDestory()

3. 下列选项中错误的是（　　）。

A. 采用 startService() 方法启动的服务，只能调用 Context.stopService() 方法结束服务，服务结束时会调用 onDestroy() 方法

B. 使用 startService() 方法启动服务，调用者与服务绑定在一起，调用者一旦退出，服务也就终止

C. 使用 bindService() 方法启动服务，调用者与服务绑定在一起，调用者一旦退出，服务也就终止

D. 如果使用 startService() 方法启动服务，在服务还未被创建时，系统会先调用服务的

onCreate() 方法，接着调用 onStart() 方法。如果调用 onStart() 方法之前服务已经被创建，多次调用 startService() 方法并不会导致多次创建服务，但会导致多次调用 onStart() 方法

4. 安卓系统中使进程之间能进行数据交换的协议是（　　）。

A. AIDL 服务　　B. TCP 协议　　C. socket 协议　　D. Http 协议

5. 下列选项正确的是（　　）。

A. 当客户端访问 Service 时，Andriod 是直接返回 Service 给客户端的

B. Andriod 的 AIDL 远程实现的接口类就是 IBinder 类

C. 本地 Service 的 onBind() 方法不会直接把 IBinder 对象本身传给 onServiceConnected 方法的第二个参数

D. AIDL 定义接口的源代码必须以 .aidl 结尾

二、填空题

1. Service 的启动有两种方式：________ 和 ________。

2. Service 的生命周期并不像 Activity 那么复杂，它只继承了 ________、________、________ 三个方法。

3. 开发一个简单的音乐播放的应用程序，启动该服务是属于 ________。

4. Service 的远程通信，也是跟其他 ________ 的通信。

5. ________ 和 ________ 同被称为 Android 的基本组件。

三、判断题

1. Service（服务）是一个没有用户界面的在后台运行执行耗时操作的应用组件。（　　）

2. 在其他进程中启动当前 Service，这是本地服务通信。（　　）

3. 绑定的 Service 只有当应用组件绑定后才能运行。（　　）

4. 停止本地服务通信，可以通过 onUnBind() 来停止。（　　）

5. Service 只存在于后台服务。（　　）

四、简答题

1. Service 有哪些种类？

2. Service 与 Thread 的区别？

五、上机题

编写一个程序实现短信拦截。

项目九　广播接收者

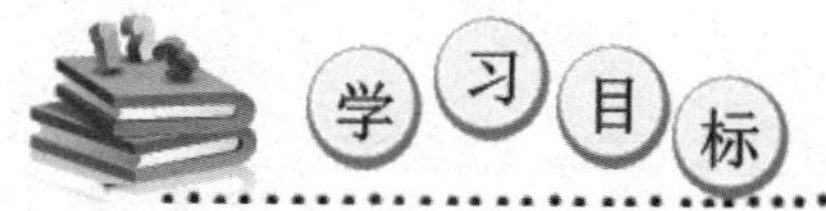

通过“闹钟系统”在应用程序之间的广播监听，学习广播的发送、监听、保存信息的相关知识和使用方法。在项目实现过程中：

- 了解广播接收者组件的作用和意义；
- 掌握使用广播发送信息；
- 掌握使用监听广播；
- 掌握保存广播相关信息。

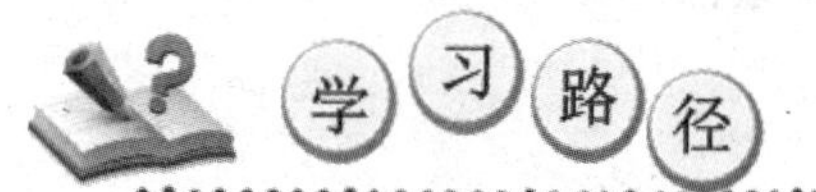

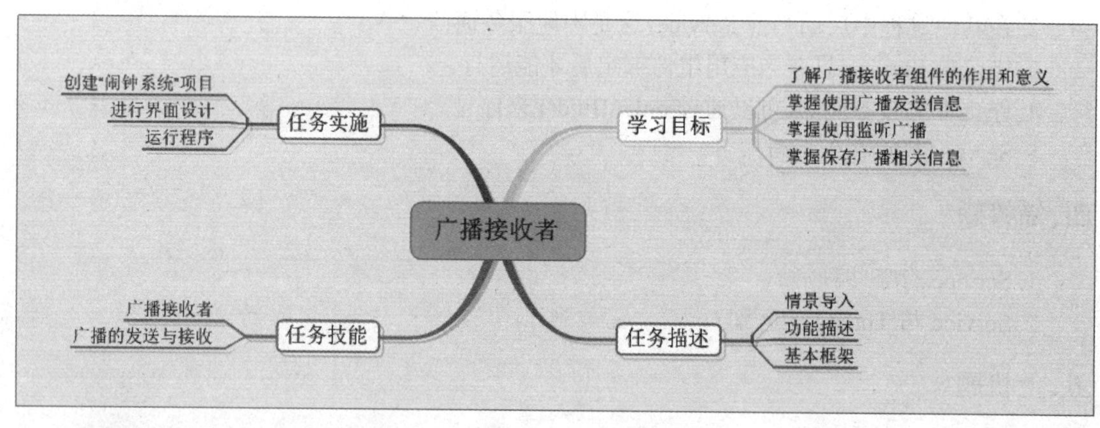

【情景导入】

智能手机已成为人们生活工作中存储发送信息文件的重要工具，而广播是一种广泛运用在应用程序之间传输信息的机制，适用于信息的发送、收和保存。本次任务主要学习使用广播接收者对信息进行接收和发送，使读者对信息的发送与存储有一定的了解。

【功能描述】

本项目将设计一款利用广播进行监听的“闹钟”软件。

- 使用线性布局技术来设计闹钟系统界面；
- 主界面点击“设置闹钟”进行时间的设置，并将设置的时间显示到界面；
- 当前时间与设置时间一致时，弹出对话框进行提示；
- 用户可点击“关闭”，将闹钟关闭；
- 在主界面中，可点击“移除闹钟”，将已设置的闹钟删除；
- 当用户退出应用程序时，会弹出“确认退出”提示框，点击“确定”按钮则退出，点击“取消”按钮则回到主界面。

【基本框架】

基本框架如图 9.1 所示，将框架图转换成的效果如图 9.2 所示。

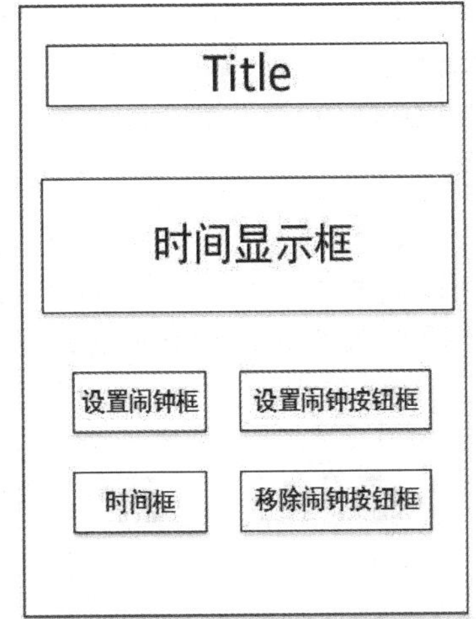

图 9.1　闹钟系统主界面

图 9.2　闹钟系统主界面

技能点 1　广播接收者

1　广播接收者简介

BroadcastReceiver 是 Android 系统的四大组件之一，BroadcastReceiver 监听的事件源是 Android 应用中的其他组件。如 startService() 方法启动的 Service 之间的通信，就可以借助 BroadcastReceiver 来实现。广播流程图如图 9.3 所示。

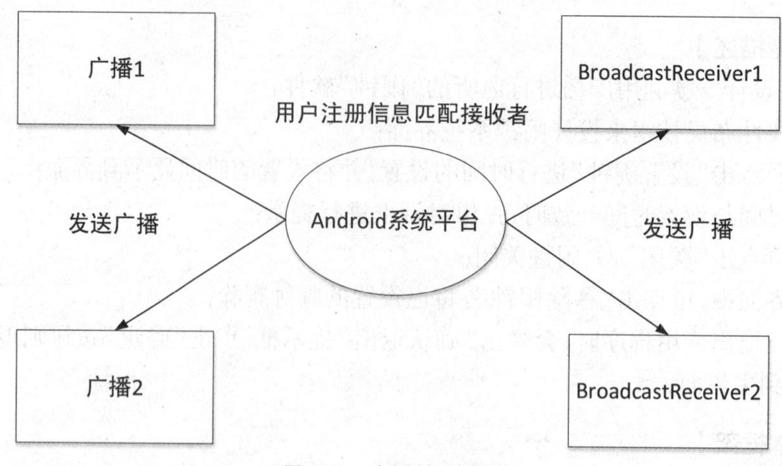

图 9.3 广播的流程图

BroadcastReceiver 用于接收程序所发出的 Broadcast Intent，与应用程序启动 Activity、Service 一样，需要两步启动：创建需要启动的 BroadcastReceiver 的 Intent；调用 Context 的 sendBroadcast() 方法或 sendOrderedBroadcast() 方法启动指定的 BroadcastReceiver。

2 广播接收者注册方法

实现 BroadcastReceiver 的方法只需重写 BroadcastReceiver 的 onReceive(Context context,Intent intent) 方法即可。实现 BroadcastReceiver 后，指定该 BroadcastReceiver 所匹配的 Intent。以下有两种方法，分别是动态注册与静态注册。

- 使用代码进行动态注册，调用 BroadcastReceiver 类的 Context 的 registerReceiver() 方法，动态注册的特点是，在代码中进行注册后，当应用程序关闭后，就不再进行监听。具体实现方法如下所示：

```
// 实现动态注册功能
MyReceiver receiver = new MyReceiver();
IntentFilter filter = new IntentFilter();
filter.addAction("android.intent.action.MY_BROADCAST");
registerReceiver(receiver, filter); // 注册
```

- 静态注册是在 AndroidManifest.xml 文件中配置，注册常驻是静态注册的特点，不论该应用是否处于活动状态，如有广播传来，将会被系统调用自动运行。具体实现方法如下所示：

```
<!—静态注册   广播接收者名称 -->
<receiver android:name=" ">
<intent-filter>
<!-- Intent-filter 过滤条件 -->
<category android:name=" ">
</intent-filter>
```

```
</receiver>
```

3　广播事件的执行

每次执行系统 Broadcast(广播)事件，系统就会创建对应的 BroadcastReceiver 实例，并自动触发它的 onReceive() 方法。onReceive() 方法执行完后，BroadcastReceiver 实例会被销毁。

若 BroadcastReceiver 的 onReceive() 方法在 10 秒内不能执行完成，系统会认为程序无响应，所以不能在 BroadcastReceiver 的 onReceive() 方法里执行一些耗时的操作，否则会弹出 ANR 对话框。

若需要根据 Broadcast 完成一个比较耗时的操作时，可以通过 Intent 启动一个 Service 完成该操作。所以 BroadcastReceiver 本身的生命周期很短，不能使用新线程去完成耗时操作，因为可能会出现线程没结束，BroadcastReceiver 就已退出的情况。BroadcastReceiver 一旦结束，此时 BroadcastReceiver 所在的进程很容易在系统需要内存时被优先杀死，因为它属于空进程（没有任何活动组件的进程）。如果所在进程被杀死，它的工作的子线程也会被杀死，采用子线程来解决有太多的问题，所以不建议使用。

BroadcastReceiver 的进程结束，即使该进程内还有用户启动的新线程，由于该进程内不包含任何活动组件，系统可能在内存不足时优先结束该进程，会导致 BroadcastReceiver 启动的子线程不能执行完毕。

广播接收者（BroadcastReceiver）其实是一种用于接收广播的 Intent，广播 Intent 的发送是通过调用 Context.sendBroadcast() 方案、Context.sendOrderedBroadcast() 方法实现。订阅了此 Intent 的多个广播接收者都可以接收此广播。要实现一个广播接收者方法步骤如下。

（1）继承 BroadcastReceiver，并重写 onReceive() 方法。

```
// 继承 BroadcastReceiver
public class IncomingSMSReceiver extends BroadcastReceiver {
@Override public void onReceive(Context context, Intent intent) {
}
}
```

（2）订阅感兴趣的广播 Intent，订阅方法有两种：

- 使用代码进行订阅；
- 在 AndroidManifest.xml 文件中的 <application> 节点里进行订阅。

```
public class IncomingSMSReceiver extends BroadcastReceiver {
@Override public void onReceive(Context context, Intent intent) {
        // 发送 Intent 启动服务，由服务来完成比较耗时的操作
        Intent service = new Intent(context, XxxService.class);
        context.startService(service);
}
```

除接收短信广播 Intent，Android 还有很多广播 Intent，如：开机启动、电池电量变化、时间已经改变等广播 Intent。接收电池电量变化广播 Intent 如下所示。

（1）在 AndroidManifest.xml 文件中的 <application> 节点里订阅此 Intent。

```
<receiver android:name=".IncomingSMSReceiver">
    <intent-filter>
        <action android:name="android.intent.action.BATTERY_CHANGED"/>
    </intent-filter>
</receiver>
```

（2）接收开机启动广播 Intent，在 AndroidManifest.xml 文件中的 <application> 节点里订阅此 Intent。

```
<receiver android:name=".IncomingSMSReceiver">
    <intent-filter>
        <action android:name="android.intent.action.BOOT_COMPLETED"/>
    </intent-filter>
</receiver>
```

（3）并且要进行权限声明。

```
<uses-permission    android:name="android.permission.RECEIVE_BOOT_COMPLET-
ED"/>
```

技能点 2　广播的发送与接收

在程序中发送广播十分简单，需要调用 Context 的 sendBroadcast(Intent intent) 方法，广播将会启动 Intent 参数所对应的 BroadcaseReceiver。以下是对如何发送 Broadcast、使用 BroacastReceiver 接收广播的介绍。

1　广播的发送

Broadcast 分为普通广播和有序广播。普通广播只能够在应用程序的内部进行传递，并且广播接收器也只能接收来自本应用程序发出的广播，这样就提高了数据传播的安全性。但普通广播无法通过静态注册的方式来接收。普通广播使用 LocalBroadcastManager 来对广播进行管理，并提供了发送广播及注册广播接收器的方法。图 9.4、图 9.5 分别为普通广播和有序广播的发送过程示意图。

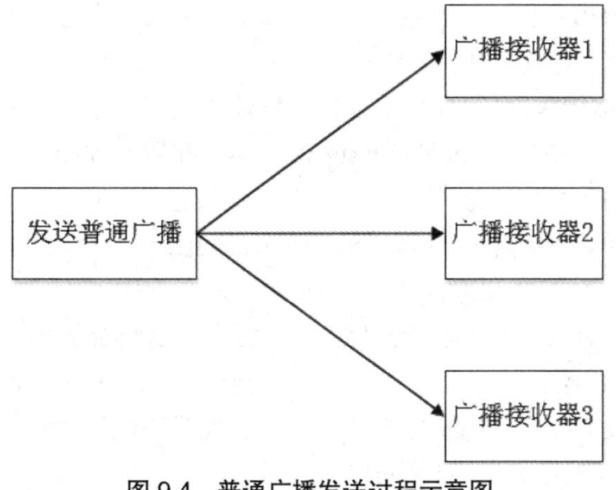

图 9.4　普通广播发送过程示意图

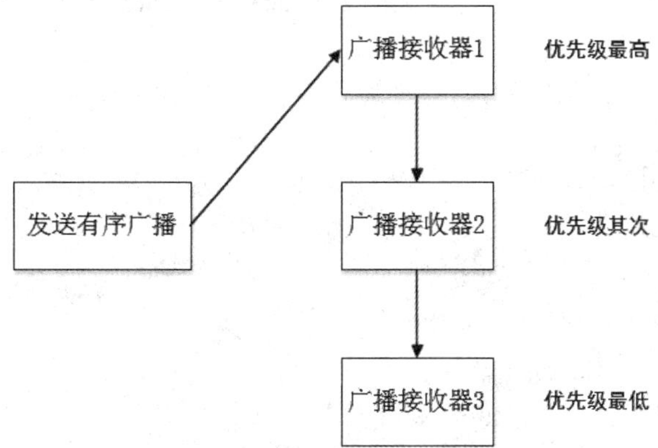

图 9.5　有序广播发送过程示意图

- 普通广播对于多个接收者来说是完全异步的，每个接收者都不需要等待便可以接收到广播，接收者之间不会有影响。接收者无法终止其他接收者的接收动作。
- 有序广播（Ordered Broadcast）比较特殊，它每次只发送到优先级较高的接收者那里，然后由优先级高的接收者再传播到优先级低的接收者那里，优先级高的接收者有能力终止这个广播。Broadcast Intent 的传播一旦终止，后面的接收者将无法接收到 Broadcast。另外，Ordered Broadcast 的接收者可以将数据传递给下一个接收者。Context 提供以下两种方法用于发送广播，如表 9.1 所示。

表 9.1　发送广播的两种方式

方法名称	说明
sendBroadcast()	发送普通广播
sendOrderedBroadcast()	发送有序广播

对于 Ordered Broadcast 而言，优先接收到 Broadcast 的接收者可以通过 setResultExtras(Bundle) 方法将处理结果存入 Broadcast 中，然后传给下一个接收者。下一个接收者通过

代码 Bundle bundle = getResultExtras(true) 可以获取上一个接收者存入的数据。

有序广播的注意事项有以下几点：

（1）有序广播被广播接收器接收时，广播接收器注册也可以不设置监听优先级即 <intent-filter android:priority="1000"> 中的 android:priority 属性不用配置。如果不设置仍然可以监听到广播，但是这样一来就是另一种监听顺序。这样可以按照监听普通的广播一样监听有序广播。

（2）BoradcastReceiver 中方法 public final boolean isOrderedBroadcast() , 可以判断当前进程正监听到的广播是否有序，如果有序返回 true, 无序返回 false。

（3）广播是否有序与广播是否有权限无关。两者也可以结合使用。

（4）多个广播接收器监听有序广播时，如果没有按照监听有序广播的形式去监听，即在注册广播接收器时不设置优先级，不同项目中的广播接收器的监听顺序就是任意的。如果在一个项目中想先收到广播，则在清单文件中就要先注册。

（5）系统收到短信时，发出的 Broadcast 就是 Ordered Broadcast。如果先实现阻止用户收到短信，就可以通过设置优先级，让自定义的 BoradcastReceiver 先获取到 Boradcast，然后终止 Boradcast。

发送自定义的无序广播的过程，首先要发送广播，代码如下：

```
public void startBroadcast(View view){
        // 开启广播
        // 创建一个意图对象
        Intent intent = new Intent();
        // 指定发送广播的频道
        intent.setAction("com.example.BROADCAST");
        // 发送广播的数据
        intent.putExtra("key", " 发送无序广播 , 顺便传递的数据 ");
        // 发送
        sendBroadcast(intent);11    }
```

然后是接收广播，需要新建一个 UnorderedReceiver 类，继承 BroadcastReceiver。具体实现方法如下：

```
public class UnorderedReceiver extends BroadcastReceiver {
    @Override
    public void onReceive(Context context, Intent intent) {
        String action = intent.getAction();
        String data = intent.getStringExtra("key");
          System.out.println(" 接收到了广播 ,action:"+ action +",data:"+data);
  // 接受到了广播 ,action:com.example.BROADCAST,data: 发送无序广播 , 顺便传
递的数据
    }}
```

2　广播的接收

BroadcastReceiver 的一个重要用途就是接收系统广播。当应用需要在系统特定时刻执行某些操作，就可以通过监听系统广播来实现。Android 的大量系统时间都会对外发送普通广播。Android 常见的广播 Action 常量如表 9.2 所示。

表 9.2　系统广播 Action 常量说明

变量名称	说明
ACTION_TIME_CHANGED	系统时间被改变
ACTION_DATA_ CHANGED	系统日期被改变
ACTION_TIMEZONE_ CHANGED	系统时区被改变
ACTION_BOOT_COMPLETED	系统启动完成
ACTION_PACKAGE_ADDED	系统添加包
ACTION_ PACKAGE_CHANGED	系统的包改变
ACTION_ PACKAGE_REMOVED	系统的包被删除
ACTION_ PACKAGE_RESTARTED	系统的包被重启
ACTION_ PACKAGE_DATA_CLEARED	系统的包数据被清空
ACTION_BATTERY_CHANGED	电池电量改变
ACTION_BATTERY_LOW	电池电量低
ACTION_POWER_CONNECTED	系统连接电源
ACTION_POWER_DISCONNECTED	系统与电源断开
ACTION_SHUTDOWN	系统被关闭

接收系统的广播首先要新建一个类继承 BroadcastReceiver。以监听 SD 卡状态的广播接收者为例，具体实现方法如下：

```
public class SdCardBroadcastReceiver extends BroadcastReceiver {
    @Override
    public void onReceive(Context context, Intent intent) {
        String action = intent.getAction();
     if("android.intent.action.MEDIA_MOUNTED".equals(action)){
            System.out.println("SD 卡已挂载 ");
        }else if("android.intent.action.MEDIA_UNMOUNTED".equals(action)){
            System.out.println("sd 卡已卸载 ");
        }
    }
}
```

然后在清单文件中注册，具体实现方法如下。

最后在清单文件中添加意图过滤器，action 里写监听的内容。具体实现方法如下：

```
    <!-- 相当于装电池 -->
            <receiver android:name="com.example.sdbroadcast.SdCardBroadcastReceiver">
    <!-- 相当于调频道 -->
                <intent-filter>
                    <action android:name="android.intent.action.MEDIA_MOUNTED"/>
                    <action    android:name="android.intent.action.MEDIA_UNMOUNT-
ED"/>
                    <data android:scheme="file"/>
                </intent-filter>
            </receiver>
```

拓展：想了解或学习广播接收者使用方法，可扫描下方二维码，获取更多信息。

第一步：在 Eclipse 中创建一个 Android 工程，命名为“闹钟系统”，并设计界面。如图 9.2 所示。

第二步：在 src 文件夹下建立 MainActivity.java 文件并实现界面初始化。

第三步：点击“设置闹钟”按钮，实现闹钟设置功能并实现闹钟监听功能。具体如代码 CORE0901 所示。

代码 CORE0901：按钮点击事件

```
/**
        * 1 此处填写闹钟时间设置代码
                */
mSet.setOnClickListener(new View.OnClickListener() {
public void onClick(View v) {
        mCalendar.setTimeInMillis(System.currentTimeMillis());
        int mHour = mCalendar.get(Calendar.HOUR_OF_DAY);// 小时
```

```
            int mMinute = mCalendar.get(Calendar.MINUTE);// 分钟
            new TimePickerDialog(ClockDemo.this,
           new TimePickerDialog.OnTimeSetListener() {
        public void onTimeSet(TimePicker view,
                    int hourOfDay, int minute) {
                            mCalendar.setTimeInMillis(System.
currentTimeMillis());
                            mCalendar.set(Calendar.HOUR_OF_DAY, hourOfDay);
                            mCalendar.set(Calendar.MINUTE, minute);
                            mCalendar.set(Calendar.SECOND, 0);
                            mCalendar.set(Calendar.MILLISECOND, 0);
                            ObjectPool.mAlarmHelper.openAlarm(32,"ddd",
                    "ffff", mCalendar.getTimeInMillis());
                    time.setText(" 时间: "+hourOfDay+":"+minute);
                                   }
           }, mHour, mMinute, true).show();
                            }
                    });
   /**
           * 2 此处填写监听闹钟代码
                    */
   public void openAlarm(int id, String title, String content, long time) {
                    Intent intent = new Intent();
                    intent.putExtra("_id", id);
                    intent.putExtra("title", title);
                    intent.putExtra("content", content);
                    intent.setClass(c, CallAlarm.class);
                    PendingIntent pi = PendingIntent.getBroadcast(c, id, intent,

                    PendingIntent.FLAG_UPDATE_CURRENT);
                    mAlarmManager.set(AlarmManager.RTC_WAKEUP, time, pi);
           }
   // 闹钟监听
   public class CallAlarm extends BroadcastReceiver {
           @Override
           public void onReceive(Context context, Intent intent) {
                   intent.setClass(context, AlarmAlert.class);
```

```
                    intent.addFlags(Intent.FLAG_ACTIVITY_NEW_TASK);
                    context.startActivity(intent);
            }
    }
    // 闹钟提示
    public class AlarmAlert extends Activity {
            @Override
            protected void onCreate(Bundle savedInstanceState) {
                    super.onCreate(savedInstanceState);
                    new AlertDialog.Builder(AlarmAlert.this)
                                    .setIcon(R.drawable.clock)
                                    .setTitle(" 闹钟 ")
                                    .setMessage(" 起床了!! ")
                                    .setPositiveButton(" 关闭 ",
                    new DialogInterface.OnClickListener() {
                    public void onClick(DialogInterface dialog,int whichButton) {
            finish(); }
                    }).show();}}
```

第四步：点击“移除闹钟”按钮，实现闹钟移除功能并实现闹钟监听功能。具体如代码 CORE0902 所示。

代码 CORE0902：移除闹钟

```
/**
        * 3 此处填写移除闹钟代码
                */
cSet.setOnClickListener(new OnClickListener() {
                @Override
                public void onClick(View arg0) {
                    // TODO Auto-generated method stub
                    ObjectPool.mAlarmHelper.closeAlarm(32, "ddd", "ffff");
                    time.setText(" 时间：");
                }
        });
// 关闭闹钟
public void closeAlarm(int id, String title, String content) {
        Intent intent = new Intent();
        intent.putExtra("_id", id);
```

```
            intent.putExtra("title", title);
            intent.putExtra("content", content);
            intent.setClass(c, CallAlarm.class);
            PendingIntent pi = PendingIntent.getBroadcast(c, id, intent, 0);
            mAlarmManager.cancel(pi);
    }
```

第五步:实现退出程序。显示确认退出提示框,具体如代码 CORE0903 所示。

代码 CORE0903:退出确定提示框

```
/**
            *4 此处填写闹钟软件退出提示代码
              */
@Override
    public boolean onKeyDown(int keyCode, KeyEvent event) {
// 判断是否点击退出
        if (keyCode == KeyEvent.KEYCODE_BACK) {
            showBackDialog();
            return true;
        }
        return super.onKeyDown(keyCode, event);
    }
    public void showBackDialog() {
        final AlertDialog.Builder builder = new AlertDialog.Builder(this);
        builder.setTitle(" 提示 ")// 设置标题

                .setIcon(R.drawable.icon) // 设置标题图表
                .setMessage(" 是否退出 ?") // 设置副标题
                .setPositiveButton(" 确定 ",
                        new DialogInterface.OnClickListener() {
                            public void onClick(DialogInterface dialog,
                                    int which) {
// 确认退出
                                System.exit(0);
                                android.os.Process
                                        .killProcess(android.os.Process
                                                .myPid());
                                dialog.dismiss();
```

```
                    }
                })
            .setNegativeButton(" 取消 ",
                new DialogInterface.OnClickListener() {
                    public void onClick(DialogInterface dialog,
                            int which) {
                        dialog.dismiss();
                    }
                });
        AlertDialog ad = builder.create();
        ad.show();
    }
```

第六步：运行程序，运行结果如图 9.6 所示。

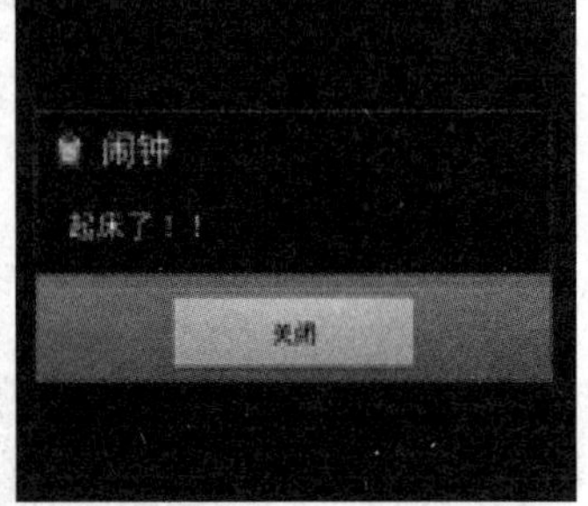

图 9.6 发送广播运行结果

【拓展目的】

熟悉并掌握使用广播接收者发送、监听、保存信息。

【拓展内容】

在“闹钟系统”的基础上实现闹铃功能，界面如图 9.3 所示。

【拓展步骤】

（1）设计思路：根据闹钟提示播放闹铃音乐。

（2）开启和关闭闹铃功能，具体如代码 CORE0904 所示。

代码 CORE0904：闹铃播放

```
/**
*1 此处添加铃声播放代码
*/
player =new  MediaPlayer();
// 创建播放器添加播放文件
        player =MediaPlayer.create(this, R.raw.a);
        player.setLooping(false);// 设置播放是否循环
        player.start();// 开始播放
// 播放通知显示闹钟提示框
        new AlertDialog.Builder(AlarmAlert.this)
                .setIcon(R.drawable.clock)
                .setTitle(" 闹钟 ")
                .setMessage(" 起床了！！ ")
                .setPositiveButton(" 关闭 ",
                        new DialogInterface.OnClickListener() {
                            public void onClick(DialogInterface dialog,
                                    int whichButton) {
                                player.stop();
                                finish();
                            }
                        }).show();
```

本项目主要介绍了 BroadcastReceiver 以及手机监听的使用方法。系统提供的 Service 和 BroadcastReceiver 的结合使用是本项目的学习难点。二者的使用都涉及权限的添加，需要读者重点掌握。通过对本项目的学习可以更加清楚地了解广播类型以及手机监听的使用方法，掌握权限添加的基本流程，提高对广播接收者的认知度。

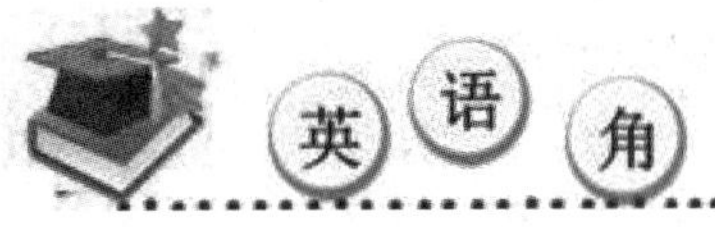

receiver	接收者
broadcast	广播
state	状态

sleep　　睡眠
update　　更新
while　　当……时
system　　系统
width　　宽
label　　标签
method　　方法

一、选择题

1.Android 开发的四大组件不包括（　　）。

A．Activity　　B. Service　　C. Broadcast Receiver　　D. Intent

2. 下列普通广播的特点中不包括（　　）。

A. 使用 onReceive() 方法获取广播消息

B. 有两种注册方式

C. 需要获取手机功能权限

D. 随开随停

3. 下面在 AndroidManifest.xml 文件中注册 BroadcastReceiver 方式正确的（　　）。

A．
```
<receiver android:name="NewBroad">
    <intent-filter>
        <action  android:name="android.provider.action.NewBroad"/>
        <action>
    </intent-filter>
</receiver>
```

B．
```
<receiver android:name="NewBroad">
        <intent-filter>
      android:name="android.provider.action.NewBroad"/>
        </intent-filter>
    </receiver>
```

C．
```
<receiver android:name="NewBroad">
    <action  android:name="android.provider.action.NewBroad"/>
    <action>
</receiver>
```

D．
```
<intent-filter>
    <receiver android:name="NewBroad">
        <action android:name="android.provider.action.NewBroad"/>
```

<action>

</receiver>

</intent-filter>

4. 下列关于 BroadcastReceiver 的说法不正确的是（　　）。

A. 是用来接收广播 Intent 的

B. 一个广播 Intent 只能被一个订阅了此广播的 BroadcastReceiver 所接收

C. 对有序广播，系统会根据接收者声明的优先级别按顺序逐个执行接收者

D. 接收者声明的优先级别在 android:priority 属性中声明，数值越大优先级别越高

5. 关于 Intent 对象说法错误的是（　　）。

A. 在 Android 中，Intent 对象是用来传递信息的

B. Intent 对象可以把值传递给广播或 Activity

C. 利用 Intent 传值时，可以传递一部分值类型

D. 利用 Intent 传值时，它的 key 值可以是对象

二、填空题

1.BroadcastReceiver 使用前需要进行注册，两种注册方法分别是 ________、________。

2.Context.sendBroadcast() 发送的广播，所有满足条件的 BroadcastReceiver 都会执行其 ________ 方法来处理响应。

3.Context.sendOrderedBroadcast 发送的有序广播，会根据 BroadcastReceiver 注册时 Intent-Filter 的 _____ 来执行 onReceive() 方法。

4. 除了接收用户发送的广播之外，BroadcastReceiver 还有一个重要的功能是 _____。

5. 如果应用需要在系统特定的时刻执行某些操作，就可以通过 ______ 来实现。

三、判断题

1. 广播接收者是一个专注于接收广播通知信息，并做出对应处理的组件。（　　）

2. 应用程序不可以进行广播，只有系统有权调用广播。（　　）

3. 广播接收者没有用户界面。然而，它们可以启动一个 Activity 来响应它们收到的信息。（　　）

4. 广播分为两种不同的类型："普通广播"和"有序广播"。（　　）

5. 有序广播的接收者可以将数据传递给下一个接收者。（　　）

四、简答题

1. BroadCastReceiver 有哪些注册方式，它们的异同是什么？

2. 广播分为哪些类型？能实现哪些手机功能？至少举 3 个实例。

五、上机题

新建项目 Time，监控手机的系统时间的变化。当系统时间被修改时，通过广播提示。

项目十　内容提供者

通过"通讯录系统"实现应用程序之间的数据提供和交换，学习 Android 内容提供者相关知识，了解（ContentProvider）和内容解析者（ContentResolver）的使用方法。在项目实现过程中：

- 掌握 ContentProvider 的功能与意义；
- 了解 ContentProvider 与 ContentResolver 的关系；
- 掌握 ContentObserver 类的作用和常用方法。

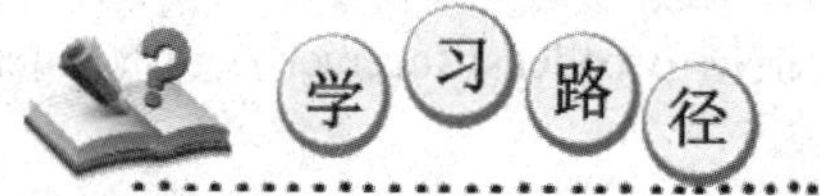

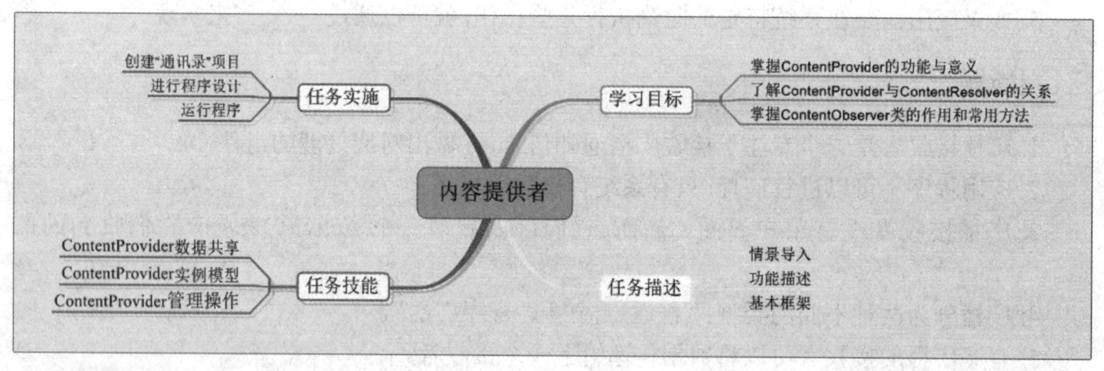

【情景导入】

内容提供者是数据库和应用程序之间的桥梁，为了能够更方便地得到用户所需信息，就需要内容提供者来获取数据。例如提供短信及联系人信息等。本任务主要使用 ContentProvider 技术，实现使用内容提供者对联系人信息的操作功能。

【功能描述】

本任务将设计一款使用 ContentProvider 共享联系人数据并且能进行修改删除的程序。

- 使用线性布局技术来设计通讯录系统界面；
- 主界面，点击“添加联系人”按钮，获取输入的联系人姓名，电话号；
- 创建数据库，将得到的信息添加到数据库中；
- 点击“查询联系人”按钮，对数据库进行查询，并将得到的数据放到 ListView 中进行显示；
- 选中 ListView 中的一条记录长按，弹出提示框“确认删除”，选中删除将数据库中信息与 ListView 中的信息进行删除，选中取消则返回主界面。

【基本框架】

基本框架如图 10.1 所示，将框架图转换成的效果如图 10.2 所示。

图 10.1　通讯录系统主界面框架图

图 10.2　通讯录系统主界面

技能点 1　ContentProvider 数据共享

1　ContentProvider 简介

Android 提供了 ContentProvider。一个程序可以通过实现一个 ContentProvider 的抽象接口将数据完全公开给其他程序，而且 ContentProvider 是以类似数据库中表的方式将数据公开。

ContentProvider 存储和检索数据，通过它可以让所有的应用程序访问到，这也是应用程序之间唯一共享数据的方法。

ContentProvider 的抽象接口，以 Uri 的形式对外界通过数据，允许其他应用程序访问或修改数据。其他程序可以使用 ContentResolver，根据 Uri 访问操作指定的数据。完整的开发一个 ContentProvider 有两个步骤：定义自己的 ContentProvider 类，该类需要继承 Android 提供的 ContentProvider 基类。在 AndroidManifest.xml 文件中注册这个 ContentProvider。注册 ContentProvider 时需要为它绑定一个 Uri。

2 Uri 简介

Uri 指定了将要操作的 ContentProvider，可以把一个 Uri 看作一个网址，把 Uri 分为三部分：

- "content://"，可以看作网址中的 "http://" 。
- 主机名或 authority，用于唯一标识这个 ContentProvider，外部应用需要根据这个标识来找到它。可以看作网址中的主机名，如 "blog.csdn.net"。
- 路径名，用来表示将要操作的数据。可以看作网址中细分的内容路径。

网址对应内容路径如图 10.3 所示。

content://contacts/people

http://blog.csdn.net/zuolongsnail

图 10.3 网址对应内容路径

3 ContentResolver 简介

应用程序的数据公开化即可被其他程序使用，可通过两种方法实现：创建一个私有的 ContentProvider；将数据添加到一个已经存在的 ContentProvider 中，前提是有相同数据类型并且有写入 ContentProvider 的权限。同时，Android 提供了 ContentResolver，外界的程序可以通过 ContentResolver 接口访问 ContentProvider 提供的数据。获取 ContentResolver 对象如表 10.1 所示。

表 10.1 Context 获取 ContentResolver 对象方法

名称	说明
getContentResolver()	获取该应用默认的 ContentResolver
insert(Uri uri,ContentValues values)	向 Uri 对应的 ContentProvider 中插入 values 对应的数据
delete(Uri uri，String where，String[] selectionArgs)	删除 Uri 对应的 ContentProvider 中 where 提交匹配的数据
update(Uri uri,ContentValues values, String where,String[] selectionArgs)	更新 Uri 对应的 ContentProvider 中 where 提交匹配的数据

续表

名称	说明
query(Uri uri,String[] projection,String selection,String[] selectionArgs,String sort Order)	查询 Uri 对应的 ContentProvider 中 where 提交匹配的数据

技能点 2　ContentProvider 实例模型

ContentProvider 实例模型介绍了 ContentProvider 与 ContentResolver 和 Uri 的关系，以及 ContentObserver 的调用方法和 ContentProvider 子类的实现方法。

1　ContentProvider、ContentResolver 和 Uri 的关系

当多个应用程序同时通过 ContentResolver 来操作 ContentProvider 提供的数据时，ContentResolver 调用的数据操作将会交给同一个 ContentProvider 处理。ContentResolver，Uri 与 ContentProvider 的关系如图 10.4 所示。

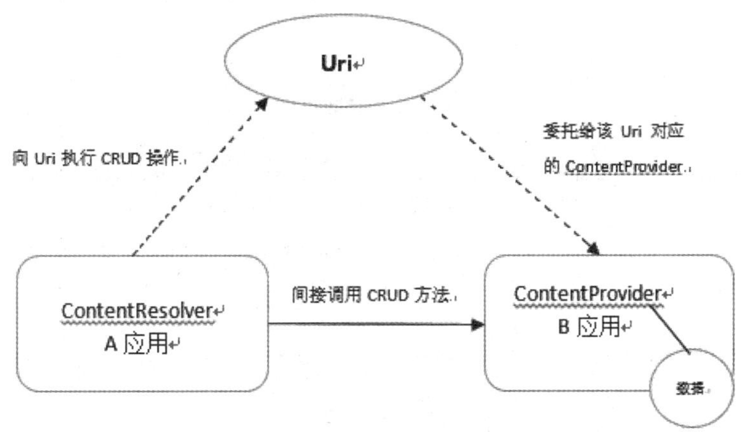

图 10.4　ContentResolverm、Uri 与 ContentProvider 的关系图

2　ContentObserver 调用方法

ContentObserver 目的是观察特定 Uri 引起的数据库的变化，随着变化做出相应的处理，当 ContentObserver 所观察的 Uri 发生变化时，就会触发 ContentObserver。触发器有两种，分别为表触发器、行触发器。ContentObserver 与 ContentProvider 和 ContentResolver 的关系如图 10.5 所示。

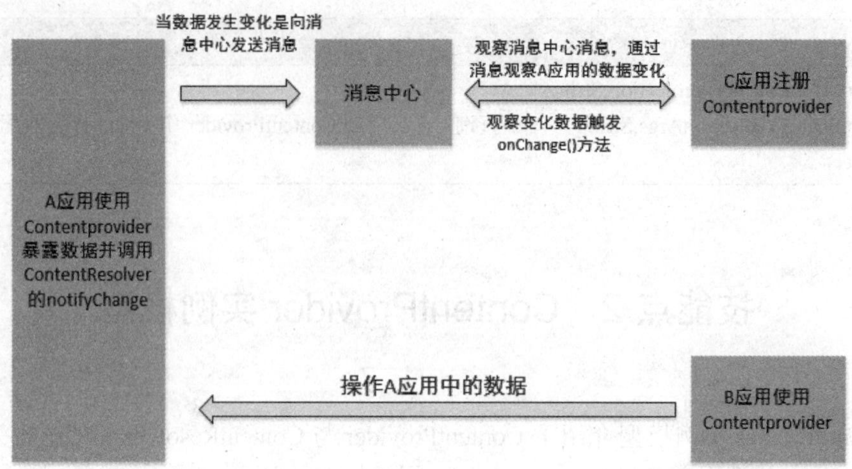

图 10.5　ContentObserver 与 ContentProvider 和 ContentResolver 的关系

ContentObserver 的构造函数中有一个参数 Handler，原因在于 ContentObserver 内部使用了一个实现 Runnable 接口的内部类，该类是 NotificationRunnable 类，用来操作数据库内容。

使用 Handler 发送消息。注册 ContentObserver 的方法是：getContentResolver().registerContentObserver(uri, notifyForDescendents, observer)。该方法中的 3 个参数说明如表 10.2 所示。

表 10.2　参数说明

名称	说明
Uri	该监听器所监听的 ContentProvider 的 Uri
notifyForDescendents	参数设为 true，监听所有与此 Uri 相关的 Uri。参数设为 false，监听特殊的 Uri
observer	监听器实例

构造方法：

```
public void ContentObserver(Handler handler)
// 解释：所有 ContentObserver 的派生类都需要调用这个构造方法
// 参数：handler：Handler 对象。可以是主线程 Handler，也可以是任何 Handler 对象
```

常用方法：

```
void onChange(boolean selfChange)
    // 功能：观察到的 Uri 发生变化时，回调该方法去处理该变化。
    // 参数：selfChange 回调后，它返回的值一般为 false 。
```

观察特定 Uri 步骤如下：

（1）创建特定的 ContentObserver 派生类，实现必须重载父类构造方法，用重载 onChange() 方法去处理所有回调后的功能实现。

（2）使用 context.getContentResolver() 获得 ContentResolver 对象，然后调用 registerContentObserver() 方法去注册内容观察者。

注意：这里需要手动地调用 unregisterContentObserver() 取消注册 ContentObserver。

3　开发 ContentProvider 子类

开发 ContentProvider 需要实现 query()、insert()、update() 和 delete() 的方法，其实质是调用指定 Uri 对应的 ContentProvider 增、删、改、查方法。ContentProvider 子类一定要在 AndroidManifest.xml 文件中注册，指定属性 android:authorities 供其他应用程序调用。ContentProvider 数据操作相关方法如表 10.3 所示。

表 10.3　ContentProvider 数据操作相关方法

名称	说明
delete(Uri uri,String selection,String[] selectionArgs)	删除一行或多行数据
insert(Uri uri,ContentValues values)	插入一行数据
query(Uri uri,String[] projection,String selection,String[] selectionArgs,String sortOrder)	查询数据
update(Uri uri,ContentValues values,String selection,String[] selectionArgs)	更新一个或多个数据

向 Android 系统中注册 ContentProvider 只要在 <application…/> 元素下添加如下子元素：具体实现方法如下所示：

```
<!—下面配置中 name 属性指定 ContentProvider 类，authorities 就相当于为该 ContentProvider 指定域名 -->
<provider android:name=".DictProvider"
    android:authorities="org.crazyit.providers.dictprovider"
    android:exported="true"/>
```

通过以上配置文件注册 DictProvider 后，其他应用程序就可以通过 Uri 访问 DictProvider 所展示的数据了。

技能点 3　ContentProvider 管理操作

Android 系统本身提供了大量的 ContentProvider，开发者开发的 Android 应用也可通过

ContentResolver 来调用系统提供的 query()、insert()、update() 和 delete() 方法对数据进行管理。

1 ContentProvider 管理联系人

Android 系统提供了 Contacts 应用程序来管理联系人，并且提供了大量的 ContentProvider，允许其他应用程序 ContentProvider 管理联系人数据。手机添加联系人时通过 ContentProvider 管理联系人实现，下面是 Android 系统用于管理联系人的 ContentProvider 的 Uri。

- ContactsContent.Cintacts.CONTENT_URI: 管理联系人 Uri;
- ContactsContent.ComminDataKinds.Phone.CONTENT_Uri: 管理人电话 Uri;
- ContactsContent.ComminDataKinds.Email. CONTENT_Uri: 管理联系人 Email 的 Uri。

2 ContentProvider 管理多媒体内容

Android 的很多 Uri 都是固定的，如 Android 提供 Camera 程序支持拍照拍摄、用户拍照、摄像等。Android 为这些多媒体提供的 ContentProvider 的 Uri 如表 10.4 所示。

表 10.4 多媒体 Uri 介绍

Uri	含义
MediaStore.Audio.Media.ExTERXML_CONTENT_URL	存储外部存储器的音频文件
MediaStore.Audio.Media.INTERNAL_CONTENT_URL	存储在手机内部存储器的音频文件
MediaStore.images,.Media.ExTERXML_CONTENT_URL	存储外部存储的图片文件
MediaStore. images.Media.INTERAL_CONTENT_URL	存储在手机内部存储器上的图片文件
MediaStore.Videoc.Media.ExTERXML_CONTENT_URL	存储在外部存储器的视频文件
MediaStore. Videoc. Media.INTERNAL_CONTENT_URL	存储在内部存储器的视频文件

拓展：想了解或学习更多 ContentProvider 使用方法，可扫描下方二维码，获取更多信息。

第一步：在 Eclipse 中创建一个 Android 工程，命名为“通讯录系统”，并设计界面。如图 10.2 所示。

第二步：在“通讯录系统”中创建一个 person 数据库。具体如代码 CORE1001 所示。

代码 CORE1001：点击“注册”按钮跳转

```
/**
*1 此处添加数据库及数据库表创建代码
*/
public class PersonSQLiteOpenHelper extends SQLiteOpenHelper {
    private static final String TAG = "PersonSQLiteOpenHelper";
    //数据库的构造方法，用来定义数据库的名称 数据库查询的结果集 数据库的版本
    public PersonSQLiteOpenHelper(Context context) {
        super(context, "person.db", null, 3);
    }
    // 数据库第一次被创建的时候调用的方法
    public void onCreate(SQLiteDatabase db) {
        // 初始化数据库的表结构
        db.execSQL("create table person (id integer primary key autoincrement, name
varchar(20), number varchar(20)) ");
    }
    // 当数据库的版本号发生变化的时候 ( 增加的时候 ) 调用
    public void onUpgrade(SQLiteDatabase db, int oldVersion, int newVersion) {
        Log.i(TAG," 数据库的版本变化了 ...");
    }
}
```

第三步：在 src 文件夹下建立 MainActivity.java 文件并编写“添加联系人”按钮单击事件，获取输入的联系人姓名，电话号码相关代码，并将信息添加到 person 数据库中。具体如代码 CORE1002 所示。

代码 CORE1002: 添加联系人信息

```
/**
*2 此处添加联系人添加代码
*/
btn_insert.setOnClickListener(new OnClickListener() {
            @Override
            public void onClick(View arg0) {
                // TODO Auto-generated method stub
                new Thread() {
                    public void run() {
                        str_name = et_name.getText().toString().trim();
                        str_phone = et_phone.getText().toString().trim();
```

```
                        if (str_name.equals("") || str_phone.equals("")) {
                            runOnUiThread(new Runnable() {
                                public void run() {
                                    Toast.makeText(MainActivity.this,
                                            " 请将信息补充完整 ", 0).show();
                                }
                            });
                        } else {
                            // 添加数据
                            AddData(str_name, str_phone);
                            runOnUiThread(new Runnable() {
                                public void run() {
                                    Toast.makeText(MainActivity.this, " 添加
成功 ", 0)
                                            .show();
                                }
                            });
                        }
                    };
                }.start();
            }
        });
    // 添加
    public void AddData(String str_name2, String str_phone2) {

        dao = new PersonDao2(this);
                Random random = new Random();
            dao.add(str_name2, str_phone2, random.nextInt(5000));
        }
    public class PersonDao2 {
        private PersonSQLiteOpenHelper helper;
        // 在构造方法里面完成 helper 的初始化
        public PersonDao2(Context context){
            helper = new PersonSQLiteOpenHelper(context);
        }
        // 添加一条记录到数据库
        public long add(String name,String number,int money){
            SQLiteDatabase db = helper.getWritableDatabase();
```

```
            ContentValues values = new ContentValues();
            values.put("name", name);
            values.put("number", number);
            long id = db.insert("person", null, values);
            db.close();
            return id;
        }
}
```

第四步：编写“查询联系人”按钮单击事件，将数据库中的信息显示到 ListView 中。具体如代码 CORE1003 所示。

代码 CORE1003：ListView 显示联系人信息

```
/**
*3 此处添加显示联系人代码
*/
btn_updata.setOnClickListener(new OnClickListener() {
            @Override
            public void onClick(View arg0) {
                // TODO Auto-generated method stub
                getPersons();
                // 如果查询到数据 则向 UI 线程发送消息
                System.out.println(persons.size()+"-------------");
                if (persons.size() > 0) {
                    handler.sendEmptyMessage(100);
                }
            }
        });
private void getPersons() {
        // 首先要获取查询的 uri
        String url = "content://cn.itcast.contentprovider.personprovider/query";
        Uri uri = Uri.parse(url);
        // 获取 ContentResolver 对象 这个对象的使用后面会详细讲解
        ContentResolver contentResolver = getContentResolver();
        // 利用 ContentResolver 对象查询数据得到一个 Cursor 对象
        Cursor cursor = contentResolver.query(uri, null, null, null, null);
        persons = new ArrayList<Person>();
        // 如果 cursor 为空立即结束该方法
```

```
            if (cursor == null) {
                return;
            }
            while (cursor.moveToNext()) {
                int id = cursor.getInt(cursor.getColumnIndex("id"));
                String name = cursor.getString(cursor.getColumnIndex("name"));
                String number = cursor.getString(cursor.getColumnIndex("number"));
                Person p = new Person(id, name, number);
                persons.add(p);
            }
            cursor.close();
        }
```

第五步：选中 ListView 中单条数据长按，弹出确认删除提示框，点击“确定”或“取消”按钮，对该条数据进行删除，并且删除在数据库中对应的信息。具体如代码 CORE1004 所示。

代码 CORE1004：对 ListView 中联系人信息进行删除操作

```
    /**
    *4 此处添加删除联系人代码
    */

    lv.setOnItemLongClickListener(new OnItemLongClickListener() {
                @Override
                public boolean onItemLongClick(final AdapterView<?> arg0, final View
view,
                        final int postion, long arg3) {
                    // TODO Auto-generated method stub
                    final TextView text=new TextView(MainActivity.this);
                    text.setText(" 您确认删除信息吗？ ");
                    text.setTextSize(23);
                    AlertDialog.Builder  builder=new  AlertDialog.Builder(MainActivi-
ty.this);
                    builder.setTitle(" 删除确认 ");
                    builder.setIcon(android.R.drawable.ic_dialog_info);
                    builder.setView(text);
                    builder.setPositiveButton(" 确 定 ",    new    DialogInterface.
OnClickListener() {
                            @Override
```

```
                    public void onClick(DialogInterface arg0, int arg1) {
                        // TODO Auto-generated method stub
// 移除联系人
dao.del(String.valueOf(persons.get(postion).getId()));
                        persons.remove(postion);
                        adapter.notifyDataSetChanged();
                    }
                });
                builder.setNegativeButton(" 取消 ",null);
                AlertDialog dialog = builder.create();
                    dialog.show();
                  return true;
                }
            });
        }
```

第六步 运行程序，运行结果如图 10.2、图 10.5 所示。

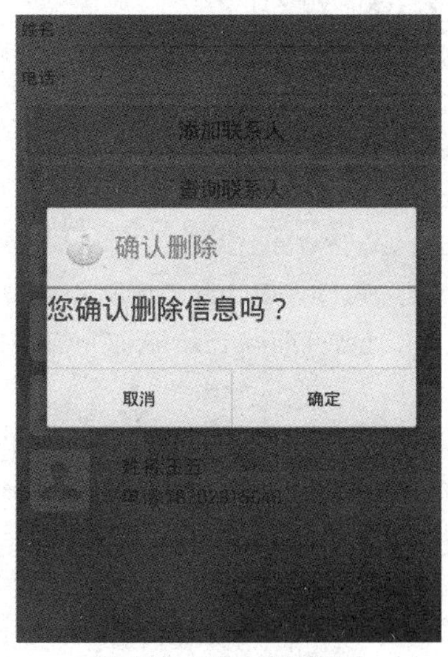

图 10.5　确认删除界面

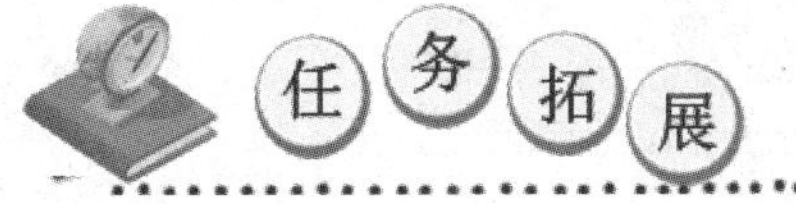

【拓展目的】

熟悉并掌握内容提供者的使用方法。

【拓展内容】

在“通讯录系统”基础上，实现长按 ListView 单条记录，弹出删除与编辑框，对数据进行编辑，效果如图 10.2 所示。

【拓展步骤】

（1）设计思路：长按 ListView 弹出编辑删除提示框。

（2）长按 ListView 中单个条目进行编辑功能，具体如代码 CORE1005 所示。

代码 CORE1005：编辑联系人信息

```
/**
*1 此处添加修改联系人信息代码
*/
    lv.setOnItemLongClickListener(new OnItemLongClickListener() {
            @Override
            public boolean onItemLongClick(final AdapterView<?> arg0,
                    final View view, final int postion, long arg3) {
                // TODO Auto-generated method stub
                final TextView text = new TextView(MainActivity.this);
                text.setText(" 请选择删除或者编辑信息 ");
                text.setTextSize(23);
                AlertDialog.Builder builder = new AlertDialog.Builder(
                        MainActivity.this);
                builder.setTitle(" 删除确认 ");
                builder.setIcon(android.R.drawable.ic_dialog_info);
                builder.setView(text);
                builder.setPositiveButton(" 删除 ",
                        new DialogInterface.OnClickListener() {
                            @Override
                            public void onClick(DialogInterface arg0, int arg1) {
                                // TODO Auto-generated method stub
                                String p = String.valueOf( persons.get(postion).
getId());
                                System.out.println(p+"------------------------");
                                dao.del(p);
                                persons.remove(postion);
                                adapter.notifyDataSetChanged();
                                System.out.println(persons.size()+"-------- 文件
大小 ---------");
                            }
```

```
                        });
    // 编辑信息
                builder.setNegativeButton(" 编辑 ",
                        new DialogInterface.OnClickListener() {
                            @Override
                            public void onClick(DialogInterface arg0, int arg1) {
                                // TODO Auto-generated method stub
                                tv_name = (TextView) view
                                        .findViewById(R.id.tv_name);
                                tv_phone = (TextView) view
                                        .findViewById(R.id.tv_phone);
                                getInfo(postion);
                            }
                        });
                AlertDialog dialog = builder.create();
                dialog.show();
                return true;
            }
        });
    private void getInfo(final int postion) {
        // TODO Auto-generated method stub
        View view = LayoutInflater.from(this).inflate(R.layout.activity_edit,
                null);
        AlertDialog.Builder builder = new AlertDialog.Builder(MainActivity.this);
        et_n = (EditText) view.findViewById(R.id.et_n);
        et_p = (EditText) view.findViewById(R.id.et_p);
        builder.setTitle(" 编辑信息 ");
        builder.setIcon(android.R.drawable.ic_dialog_info);
        builder.setView(view);
        builder.setPositiveButton(" 确定 ", new DialogInterface.OnClickListener() {
            @Override
            public void onClick(DialogInterface arg0, int arg1) {
                // TODO Auto-generated method stub
                str_n = et_n.getText().toString().trim();
                str_p = et_p.getText().toString().trim();
                tv_name.setText(" 姓名：" + str_n);
                tv_phone.setText(" 电话：" + str_p);
                String p =String.valueOf( persons.get(postion).getId());
```

```
                    System.out.println(p+"------------------------");
    // 更新数据库信息
    dao.upd(str_n, str_p, p);
                }
            });
            builder.setNegativeButton(" 取消 ", null);
            AlertDialog dialog = builder.create();
            dialog.show();
        }
```

本项目主要介绍了 Android 系统中 ContentProvider 组件的功能和用法，ContentProvider 是 Android 系统内不同程序之间进行数据交换的标准接口。通过本项目的学习需要掌握三个 API 的用法：ContentProvider、ContentResolver 和 ContentObserver，其中 ContentResolver 用于操作 ContentProvider 提供的数据，ContentObserver 用于监听 ContentProvider 的数据改变，ContentProvider 是所有 ContentProvider 组件的基类。

provider	供应者
content	内容
resolver	解决问题者
observer	观察员
message	消息
object	对象
public	公共的
throw	扔
add	增加
info	信息

一、选择题

1. 定义 ContentProvider 类的子类，下列方法不能定义的是（　　）。

A.OnStart()　B.onCreate()　C.update()　D.insert()

2. 下列知识点错的是（　　）。

A. 当外部应用需要对 ContentProvider 中的数据进行添加、删除、修改和查询操作时，可以使用 ContentResolver 类来完成，要获取 ContentResolver 对象

B. ContentProvider 可以在代码中通过 setReadPermission() 和 setWritePermission() 两个方法来设置 ContentProvider 的操作权限，也可以在配置文件中通过 android:readPermission 和 android:writePermission 属性来控制

C. ContentResolver 实例带的方法不能实现找到指定的 Contentprovider 并获取到 Contentprovider 的数据

D. Android 提供了一些主要数据类型的 Contentprovider，比如音频、视频、图片和私人通讯录等

3.Uri 指定了将要操作的 ContentProvider，并不包含的部分是（　　）。

A."content://"　B. 主机名或 authority　C. 路径名　D. 文件名

4. 要从一个 provider 中获取数据，你的应用需要对目标 provider 具有"读权限"。那么需要在（　　）文件中写入权限。

A.manifest 文件　B.Array 文件　C.string 文件　D.color 文件

5. 下列说法正确的是（　　）。

A. insert(Uri, ContentValues) 用于查询指定 Uri 的 ContentProvider，返回一个 Cursor

B. getType(Uri) 用于从指定 Uri 的 ContentProvider 中删除数据

C. onCreate() 在创建 ContentProvider 时调用

D. insert(Uri, ContentValues) 用于更新指定 Uri 的 ContentProvider 中的数据

二、填空题

1. Android 四大组件是 ________、________、________、________。

2. ContentProvider 使用继承 ContentProvider 类实现 ContentProvider 在 manifest 中注册 ________。

3. 创建一个类 NoteContentProvider，继承 ContentProvider，需要实现下面 5 个方法：________、________、________、________、________。

4. ContentProvider 三个关键类：________、________、________（表示返回结果），URI（标识 Provider, 作为 Resolver 的查询参数）。

5. 要获取操作一个 provider 的权限，应用需在自己的 ________ 文件中使用 ________ 元素。

三、判断题

1. Android 中的 ContentProvider 机制可支持在多个应用中存储和读取数据。（　　）

2. 客户端一般通过 ContentResolver（内容解析器）间接地获取内容。（　　）

3. ContentResolver 不可以与任意 ContentProvider 进行交互。（　　）

4. 如果要处理一个新的数据类型，需要定义一个新的 MIME 类型，在 ContentProvider.getType() 中返回。（　　）

5. 创建自己的 ContentProvider，不需要继承 ContentProvider 类。 ( )

四、简答题

1.ContentProvider 是什么？它的作用是什么？

2.Android 之 ContentProvider 适用场景？

五、上机题

1. 设计一个应用程序，工程名为 provider 实现数据操作：添加单击“删除”按钮，实现删除所有选中的用户，弹出删除成功提示信息，并自动刷新列表中的数据。单击“群发”按钮，实现向所有选中的用户群发短信。单击“清除”按钮，所有勾选的复选框设置为未勾选。

界面如下，在编辑框中输入信息，实现保存和删除功能。

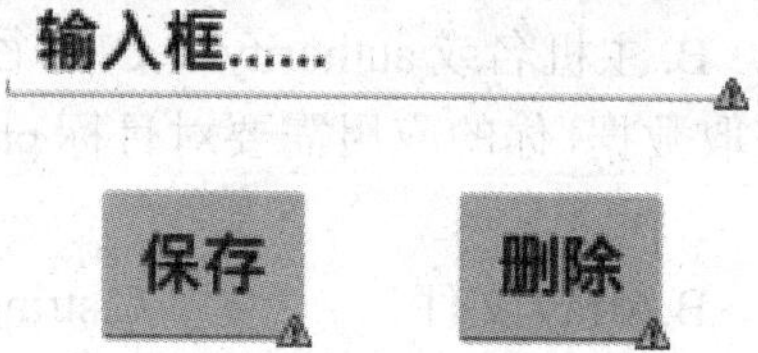

项目十一　传感器

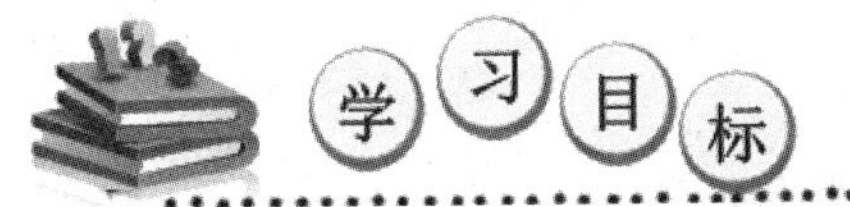

通过“摇一摇录音系统”的实现，学习加速度和重力传感器以及 Android 手机振动的相关知识，了解加速度，重力等传感使用方法。在项目实现过程中：

- 了解传感器相关知识及方法；
- 掌握 Sensor 信息检测；
- 掌握 Android 手机振动的相关方法。

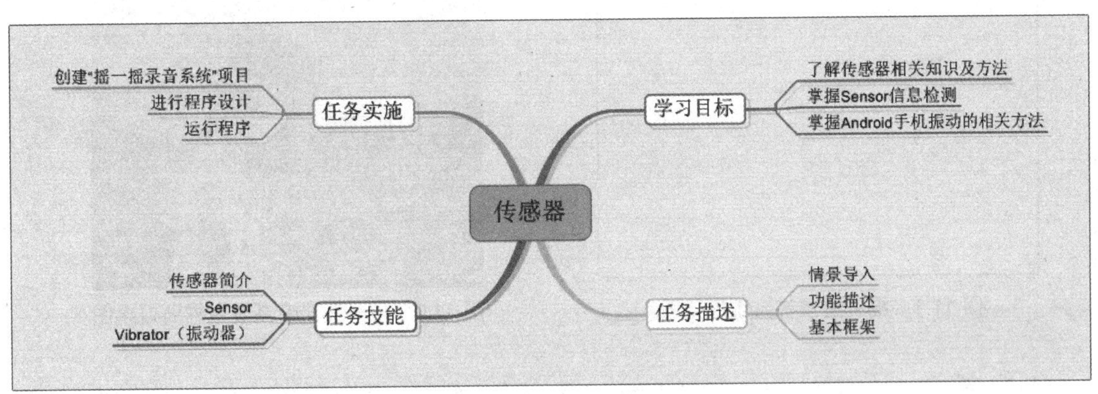

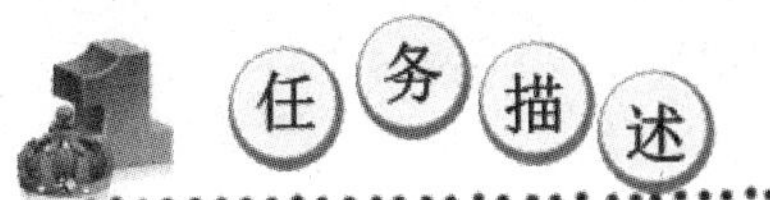

【情景导入】

越来越多的手游及应用型软件进入了人们的视野，例如赛车类手游和指南针，计步器等软件。很多人也已经进行了试用，并且对功能的实现感到新奇。本项目主要实现传感器振动录音功能。

【功能描述】

本项目将设计一款利用加速度传感器开启手机录音功能并振动提醒的“摇一摇录音系统”程序。

- 使用线性布局技术来设计登录系统界面；
- 第一次摇动手机开启录音功能；
- 第二次摇动手机保存录音；
- 点击“播放”按钮，对保存的录音进行播放；
- 点击“停止”按钮，停止播放录音。

【基本框架】

基本框架如图 11.1 所示，将框架图转换成的效果如图 11.2 所示。

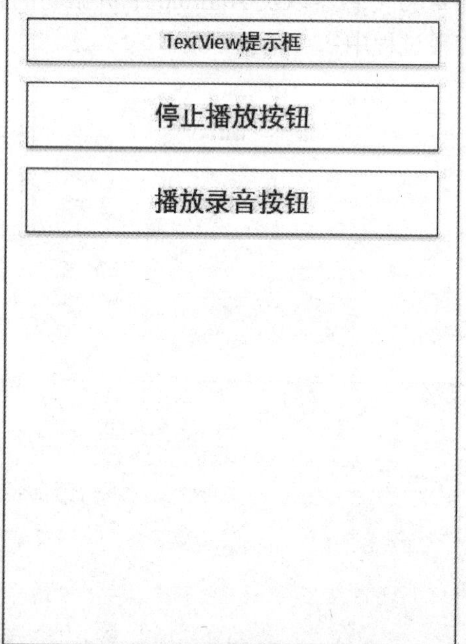

图 11.1 摇一摇录音系统框架图

图 11.2 摇一摇录音系统主界面效果图

技能点 1 传感器简介

传感器是一种监测装置，用于监测不同的数据并按照一定的规律转换成可用信息进行显示。传感器通常由敏感元件和转换元件组成，让物体拥有触觉，味觉，嗅觉等一系列感官。目前，在 Android 中提供了 11 种系统传感器供应用程序使用，如表 11.1 所示。

表 11.1　系统传感器

感应检测	说明
TYPE_ACCELEROMETER	加速度传感器
TYPE_GRAVITY	重力传感器
TYPE_AMBIENT_TEMPERATURE	温湿度传感器
TYPE_GYROSCOPE	陀螺仪传感器
TYPE_LIGHT	光感传感器
TYPE_LINEAR_ACCELERATION	线性加速度传感器
TYPE_MAGNETIC_FIELD	磁场传感器
TYPE_PRESSURE	压力传感器
TYPE_PROXIMITY	接近传感器
TYPE_ORIENTATION	方向传感器
TYPE_ROTATION_VECTOR	旋转矢量传感器

1　加速度传感器

加速度传感器用于获取手机设备的加速度状态，返回 x、y、z 三个轴的加速度值如图 11.3 所示。x 轴表示左右移动的加速度，y 轴表示前后移动的加速度，z 轴表示垂直方向的加速度，该值包含地心引力的影响。加速度传感器是众多传感器产品中比较成熟的一种。

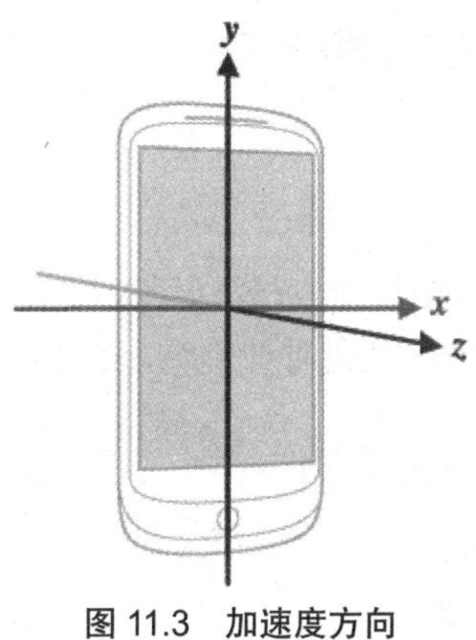

图 11.3　加速度方向

获取加速度值步骤如下：

注册加速度传感器监听事件。

```
    sensorManager.registerListener(this,
    sensorManager.getDefaultSensor(Sensor.TYPE_ACCELEROMETER),SensorManager.
SENSOR_DELAY_UI);// 注册加速度传感器监听事件
```

实时监听加速度传感器值的变化，并做出相应动作。

```
@Override
 public void onSensorChanged(SensorEvent event) {
     // TODO Auto-generated method stub
     switch (event.sensor.getType()) {
     case Sensor.TYPE_ACCELEROMETER:// 传感器类型为加速度传感器
// 判断加速度方向超出范围将进行相对应的动作
if(Math.abs(event.values[0])>17||Math.abs(event.values[1])>17||Math.abs(event.values[2])>17){
}
     break;
        default:
     break;
  }
}
```

2 重力传感器

重力传感器会返回三维向量，根据向量可得到重力的方向和强度。重力传感器的坐标系统与加速度传感器的坐标系统是相同的。

3 温度传感器

温度传感器用于获取手机所处环境的温度。温度传感将返回一个温度数据，表示手机周围温度，单位是摄氏度。

4 陀螺仪传感器

陀螺仪传感器用于感应手机的旋转速度，返回手机绕 x、y、z 这三个坐标轴的旋转速度，单位是弧度 / 秒。旋转速度为正值时代表逆时针旋转，反之为顺时针旋转。关于陀螺仪传感器的三个返回值说明。

- 表示手机绕 x 轴旋转的角速度；
- 表示手机绕 y 轴旋转的角速度，
- 表示手机绕 z 轴旋转的角速度。

5 光照传感器

光照传感器用于获取手机周围光照强度值。单位是勒克斯（Lux），其物理意义是照射到单位面积上的光通量。

6　线性加速度传感器

线性加速度传感器是加速度传感器减去重力影响获取的数据，并且其坐标系统与加速度传感器的坐标系统相同。

7　磁场传感器

磁场传感器主要用于获取手机周围磁场强度值，即使周围没有任何直接磁场，手机也处于地球磁场中。磁场会根据手机的摆放位置改变，磁场传感器会返回 x、y、z 三轴的环境磁场数据，单位为微特斯拉。

8　压力传感器

压力传感器用于获取手机所处环境的压力大小，单位是百帕斯卡（hPa）。压力传感器会返回一个数据，代表手机周围的压力大小。

9　接近传感器

接近传感器用于检测物体与手机之间的距离，单位是厘米。该传感器返回远和近两个数据，接近传感器将最大距离返回远状态，小于最大距离则返回近状态。接近传感器用于接听电话时自动关闭 LCD 屏幕以节省电量。

10　方向传感器

方向传感器用于获取手机摆放状态，可获取三个方向的角度，根据角度可确定手机的摆放状态。关于返回的三个角度说明如下：

- 表示手机顶部朝南与正北方的夹角；
- 表示手机顶部或尾部翘起的角度；
- 表示手机左侧或右侧翘起的角度。

11　旋转矢量传感器

旋转矢量传感器用于获取手机的方向，将坐标轴和角度混合运算得到的数据。传感器与轴旋转的方向相同并且得到的数据是没有单位的，使用坐标系与加速度相同。

拓展：想了解或学习更多获取传感器值的使用方法，可扫描下方二维码，获取更多信息。

技能点 2 Sensor

1 Sensor 简介

感应检测 Sensor 的硬件组件由不同的厂商提供，不同的 Sensor 设备组件所检测的事件不同。可以用 Sensor 类的相关方法检测设备所支持的 Sensor 信息，如表 11.2 所示。

表 11.2 Sensor 的相关方法

名称	说明
public float getMaximumRange()	获取 Sensor 最大值
public int getMinDelay()	获取 Sensor 的最小延迟
public String getName()	获取 Sensor 名称
public float getPower()	获取 Sensor 使用时所耗功率
public float getResolution()	获取 Sensor 的精度
public int getType()	获取 Sensor 类型
public String getVendor()	获取 Sensor 供应商信息
public int getVersion()	获取 Sensor 版本号信息

2 感应监测管理

（1）取得 SensorManager

使用感应监测 Sensor 首先获取系统服务感应设备的监测信号 Context.getSystemService (SENSER_SERVICE) 的方法获取感应监测服务。

```
mSensorManager = (SensorManager)getSystemService(SENSOR_SERVICE);
```

（2）取得感应监测 Sensor 状态的监听

```
// 在感应检测到 Sensor 的精密度有变化时被调用到。
public void onAccuracyChanged(Sensor sensor,int accuracy);
// 在感应检测到 Sensor 的值有变化时会被调用到。
public void onSensorChanged(SensorEvent event);
```

（3）实现获取感应监测 Sensor 值

实现 getSensorList() 方法获取值。

```
List<Sensor> sensors = sm.getSensorList(Sensor.TYPE_TEMPERATURE);
```

（4）实现注册 SensorListener

```
sm.regesterListener(SensorEventListener listener, Sensor sensor, int rate);
```

第一个参数表示监听 Sensor 事件，第二个参数表示 Sensor 目标种类的值，第三个参数表示延迟时间的精度和密度。

（5）取消注册

```
sm.unregisterListener(SensorEventListener listener)
```

技能点 3　Vibrator

1　Vibrator 简介

Android 手机中的振动由 Vibrator（振动器）实现。系统调用 Context 的 getSystemService() 方法，进而调用 Vibrator 相关方法如表 11.3 所示，添加相应的振动权限，便可调用 Vibrator 的方法控制手机振动。Android 振动器系统的基本层次结构如图 11.4 所示。

表 11.3　Vibrator 相关方法

方法	说明
vibrate(long milliseconds)	控制手机振动时间
vibrate(long[] pattern,int repeat)	指定手机以 pattern 的模式振动，如 pattern 为 new int[100,200,300,400]，则指 100 ms、200 ms、300 ms、400 ms 这些时间点交替启动关闭振动；repeat 指定数组中索引，从 repeat 索引开始振动循环
Boolean hasVibrator()	检测是否有振动硬件
cancel()	关闭手机振动

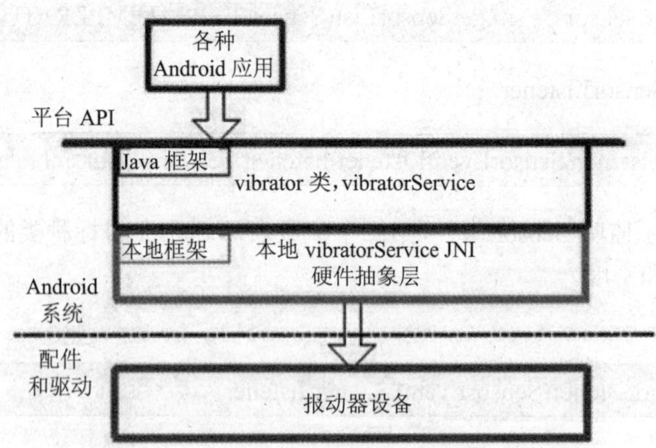

图 11.4 Android 振动器系统的基本层次结构

2 实现振动步骤

（1）获取振动服务。

```
vibrator = (Vibrator)getSystemService(Context.VIBRATOR_SERVICE)
```

（2）设置振动时间以及次数。

```
long [] pattern = {100,400,100,400};  // 停止 开启 停止 开启
vibrator.vibrate(pattern,2);           // 重复两次上面的 pattern。如果只想振动一次，
index 设为 -1
```

（3）在 AndroidManifest.xml 文件中添加手机振动所需的权限。

```
<uses-permission android:name="android.permission.VIBRATE"/>
```

第一步：在 Eclipse 中创建一个 Android 工程，命名为“摇一摇录音系统”，并设计界面。如图 11.2 所示。

第二步：在 src 文件夹中下建立 MainActivity.java 文件，并实现通过 MediaRecorder 录制音频。具体如代码 CORE1101 所示。

代码 CORE1101：MediaRecorder 录制音频

```
/**
*1 此处添加音频录制代码
```

```
    */
    mediaRecorder=new MediaRecorder();
            //2. 调用 MediaRecorder 对象的方法来设置声音来源
            mediaRecorder.setAudioSource(MediaRecorder.AudioSource.MIC);
            //3. 设置录制的音频格式
            mediaRecorder.setOutputFormat(MediaRecorder.OutputFormat.THREE_
GPP);
            //4. 设置编码格式
            mediaRecorder.setAudioEncoder(MediaRecorder.AudioEncoder.AMR_NB);
            //5. 设置保存路径
            mediaRecorder.setOutputFile
            (Environment.getExternalStorageDirectory().
                        getAbsolutePath()+"/a.mp3");
            //6. 进入准备录制的状态
            try {
                mediaRecorder.prepare();
            } catch (Exception e) {
                // TODO Auto-generated catch block
                e.printStackTrace();
            }
```

第三步：编写获取振动代码，具体如代码 CORE1102 所示。

代码 CORE1102：获取振动

```
    // 获取振动服务
    mVibrator=(Vibrator) getSystemService(VIBRATOR_SERVICE);
```

第四步：获取加速度值，摇动手机，实现录音与停止录音功能。具体如代码 CORE1103 所示。

代码 CORE1103：摇动手机，实现录音与停止录音

```
    /**
    *2 此处添加获取加速度信息实现开始与停止录音代码
    */
    @Override
        protected void onResume() {
            // TODO Auto-generated method stub
            // 传感器
```

```
            sensorManager.registerListener(this, sensorManager.getDefaultSensor(Sensor.
TYPE_ACCELEROMETER),
    SensorManager.SENSOR_DELAY_UI);
            super.onResume();
        }
    public void onSensorChanged(SensorEvent event) {
            // TODO Auto-generated method stub
            switch (event.sensor.getType()) {
                case Sensor.TYPE_ACCELEROMETER:
    if(Math.abs(event.values[0])>17||Math.abs(event.values[1])>17||Math.abs(event.val-
ues[2])>17){
                    try {
    // 判断是否摇动手机 flag==0 为摇动，flag==1 为未摇动
                        if(flag==0){
                            mVibrator.vibrate(100);// 开启振动 100ms
                            System.out.println("1111111111111111111111");
                            mediaRecorder.start();// 开始录音
                            Thread.sleep(500);// 线程延时
                            start.setText(" 开始录制中 ...");
                            System.out.println("2222222222222222");
                            Toast.makeText(MainActivity.this, " 开 始 录 制 ",
0).show();
                            flag = 1;
                        }else if(flag==1){
                            Thread.sleep(500);
                            mVibrator.vibrate(100);
                            mediaRecorder.stop();
                            start.setVisibility(View.GONE);// 隐藏
                            mediaRecorder.release();// 释放资源
                            Toast.makeText(MainActivity.this, " 录 制 完 成 ",
0).show();
                            flag = 0;
                        }
                    } catch (InterruptedException e) {
                        // TODO Auto-generated catch block
                        e.printStackTrace();
                    }
                    }
```

```
                break;
            default:
                break;
        }
    }
```

第五步：实现播放录音与停止播放。具体如代码 CORE1104 所示。

代码 CORE1104：播放与停止播放录音

```
/**
*3 此处添加播放与停止录音代码
*/
public void stop(View view){
        Toast.makeText(this, " 停止播放录音 ", 0).show();
        mediaPlayer.stop();
    }
    public void play(View view){
        try {
            // 这个是录音的存储位置和名字
            String path=Environment.getExternalStorageDirectory().
                    getAbsolutePath()+"/a.mp3";
            System.out.println(path+"---------------");
            mediaPlayer.setDataSource(path);
            mediaPlayer.prepare();
            Toast.makeText(this, " 开始播放录音 ", 0).show();
        } catch (Exception e) {
            // TODO Auto-generated catch block
            e.printStackTrace();
        }
        mediaPlayer.start();
    }
```

第六步：运行程序，结果如图 11.2 所示 。

【拓展目的】

熟悉并掌握使用加速度以及重力传感器的使用。

【拓展内容】

开发一款仿微信摇一摇软件。效果如图 11.5 所示。

图 11.5 摇一摇拓展主界面

【拓展步骤】

（1）设计思路：实现摇一摇振动，主界面实现动画。

（2）手机摇晃监听以及重力监听结果。具体如代码 CORE1105 所示。

代码 CORE1105：获取加速度值

```
/**
*1 此处填写获取重力加速度功能代码
*/
public class ShakeListener implements SensorEventListener {
    private static final int SPEED_SHRESHOLD = 2000; // 速度阈值，当摇晃速度达到这个值后产生作用
    private static final int UPTATE_INTERVAL_TIME = 70; // 两次检测的时间间隔
    private SensorManager sensorManager; // 传感器管理器
    private Sensor sensor; // 传感器
    private OnShakeListener onShakeListener; // 加速度感应监听器
    private Context mContext; // 上下文
    private long lastUpdateTime; // 上次检测时间
    // 手机上一个位置时加速度感应坐标
    private float lastX;
```

```
        private float lastY;
        private float lastZ;
        public ShakeListener(Context c) {
            // 获得监听对象
            mContext = c;
            start();
        }
        public void start() {
            // 获得传感器管理器
            sensorManager = (SensorManager) mContext
                    .getSystemService(Context.SENSOR_SERVICE);
            if (sensorManager != null) {
                // 获得加速度传感器
                sensor   =   sensorManager.getDefaultSensor(Sensor.TYPE_ACCELER-
OMETER);
            }
            // 注册
            if (sensor != null) {
                sensorManager.registerListener(this, sensor,
                        SensorManager.SENSOR_DELAY_GAME);
            }else{

                Toast.makeText(mContext, " 您的手机不支持该功能 ", 0).show();
            }
        }
        // 加速度感应器感应获得变化数据
        public void onSensorChanged(SensorEvent event) {
            long currentUpdateTime = System.currentTimeMillis();        // 当前检测时
间
            long timeInterval = currentUpdateTime - lastUpdateTime;                  //   两
次检测的时间间隔
            // 判断是否达到了检测时间间隔
            if (timeInterval < UPTATE_INTERVAL_TIME)
                return;
            // 现在的时间变成 last 时间
            lastUpdateTime = currentUpdateTime;
            // 获得 x,y,z 坐标
            float x = event.values[0];
```

```
            float y = event.values[1];
            float z = event.values[2];
            // 获得 x,y,z 的变化值
            float deltaX = x - lastX;
            float deltaY = y - lastY;
            float deltaZ = z - lastZ;
            // 将现在的坐标变成 last 坐标
            lastX = x;
            lastY = y;
            lastZ = z
            double speed = Math.sqrt(deltaX * deltaX + deltaY * deltaY + deltaZ
                        * deltaZ)
                        / timeInterval * 10000;
            // 达到速度阀值，发出提示
            if (speed >= SPEED_SHRESHOLD) {
                onShakeListener.onShake();
            }
        }
        // 摇晃监听接口
        public interface OnShakeListener {
            public void onShake();
        }
        // 停止检测
        public void stop() {
            sensorManager.unregisterListener(this);
        }
        public void onAccuracyChanged(Sensor sensor, int accuracy) {
        } // 设置重力感应监听器
        public void setOnShakeListener(OnShakeListener listener) {
            onShakeListener = listener;
        }
    }
```

（3）把 assets 目录下的声音存放在 map 中，具体如代码 CORE1106 所示。

代码 CORE1106：调用工具类方法把 assets 目录下的声音存放在 map

```
/**
*2 此处填写录音存放功能代码
```

```
*/
public class Utils {
    public static void startAnim(RelativeLayout mImgUp, RelativeLayout mImgDn) {
// 定义摇一摇动画
        AnimationSet animUp = new AnimationSet(true);
        TranslateAnimation start0 = new TranslateAnimation(
                Animation.RELATIVE_TO_SELF, 0f, Animation.RELATIVE_TO_
SELF, 0f,
                Animation.RELATIVE_TO_SELF, 0f, Animation.RELATIVE_TO_
SELF,
                -0.5f);// 位移动画效果设置
        start0.setDuration(1000);// 动画时间 1 秒
        TranslateAnimation start1 = new TranslateAnimation(
                Animation.RELATIVE_TO_SELF, 0f, Animation.RELATIVE_TO_
SELF, 0f,
                Animation.RELATIVE_TO_SELF, 0f, Animation.RELATIVE_TO_
SELF,
                +0.5f); // 位移动画效果设置
        start1.setDuration(1000); // 动画时间 1 秒
        start1.setStartOffset(1000);
        animUp.addAnimation(start0);
        animUp.addAnimation(start1);
        mImgUp.startAnimation(animUp);
        AnimationSet animDn = new AnimationSet(true);
        TranslateAnimation end0 = new TranslateAnimation(
                Animation.RELATIVE_TO_SELF, 0f, Animation.RELATIVE_TO_
SELF, 0f,
                Animation.RELATIVE_TO_SELF, 0f, Animation.RELATIVE_TO_
SELF,
                +0.5f); // 位移动画设置
        end0.setDuration(1000);
        TranslateAnimation end1 = new TranslateAnimation(
                Animation.RELATIVE_TO_SELF, 0f, Animation.RELATIVE_TO_
SELF, 0f,
                Animation.RELATIVE_TO_SELF, 0f, Animation.RELATIVE_TO_
SELF,
                -0.5f); // 位移动画设置
        end1.setDuration(1000);
```

```
            end1.setStartOffset(1000);
            animDn.addAnimation(end0);
            animDn.addAnimation(end1);
            mImgDn.startAnimation(animDn);
        }
        /**
        * 把 assets 目录下的声音资源添加到 map 中
        * */
        public static Map<Integer, Integer> loadSound(final SoundPool pool,
                final Activity context) {
            final Map<Integer, Integer> soundPoolMap = new HashMap<Integer, Inte-
ger>();
            new Thread() {
                public void run() {
                    try {
                        soundPoolMap.put(
                                0,
                                pool.load(
                        context.getAssets().openFd("sound/shake_sound_male.mp3"), 1));
                        soundPoolMap.put(
                                1,
                                pool.load(
                                        context.getAssets().openFd(
                                                "sound/shake_match.mp3"), 1));
                    } catch (IOException e) {
                        e.printStackTrace();
                    }
                }
            }.start();
            return soundPoolMap;
        }
    }
```

（4）手机摇晃监听以及重力监听，并且实现振动。具体如代码 CORE1107 所示。

代码 CORE1107：调用手机摇晃监听以及重力监听

```
/**
*3 此处填写手机重力监听，并实现振动功能代码
```

```
*/
public class ShakeActivity extends Activity {
    ShakeListener mShakeListener = null;
    Vibrator mVibrator;
    private RelativeLayout mImgUp;
    private RelativeLayout mImgDn;
    private SoundPool sndPool;
    private Map<Integer, Integer> loadSound;
    @Override
    public void onCreate(Bundle savedInstanceState) {
        super.onCreate(savedInstanceState);
        setContentView(R.layout.shake_activity);
        // 初始化数据
        init();
        // 调用工具类方法把 assets 目录下的声音存放在 map 中，返回一个 Hash-
Map
        loadSound = Utils.loadSound(sndPool, this);
        // 创建加速度监听器的对象
    }
    @Override
    protected void onResume() {
        super.onResume();
        mShakeListener = new ShakeListener(this);
        // 加速度传感器，达到速度阀值，播放动画
        mShakeListener.setOnShakeListener(new OnShakeListener() {
            public void onShake() {
                Utils.startAnim(mImgUp, mImgDn); // 开始 摇一摇手掌动画
                mShakeListener.stop();// 停止加速度传感器
                sndPool.play(loadSound.get(0), (float) 1, (float) 1, 0, 0,
                        (float) 1.2);// 摇一摇时播放 map 中存放的第一个声音
                startVibrato();// 振动
                new Handler().postDelayed(new Runnable() {
                    public void run() {
                        sndPool.play(loadSound.get(1), (float) 1, (float) 1, 0,
                                0, (float) 1.0);// 摇一摇结束后播放 map 中存
放的第二个声音
                        Toast.makeText(getApplicationContext(),
```

```
                                        " 抱歉，暂时没有找到 \n 在同一时刻摇一摇
的人。\n 再试一次吧！ ", 10).show();
                                mVibrator.cancel();// 振动关闭
                                mShakeListener.start();// 再次开始检测加速度传感器值
                            }
                        }, 2000);
                    }
                });
            }
            @Override
            protected void onPause() {
                super.onPause();
                if (mShakeListener != null) {
                    mShakeListener.stop();
                }
            }
        // 实现振动功能
            private void init() {
                mVibrator = (Vibrator) getApplication().getSystemService(
                        VIBRATOR_SERVICE);
                mImgUp = (RelativeLayout) findViewById(R.id.shakeImgUp);
                mImgDn = (RelativeLayout) findViewById(R.id.shakeImgDown);
                sndPool = new SoundPool(2, AudioManager.STREAM_MUSIC, 5);
            }
            public void startVibrato() { // 定义振动
                mVibrator.vibrate(new long[] { 500, 200, 500, 200 }, -1); // 第一个{ }里面是
节奏数组，
            }
            public void shake_activity_back(View v) { // 标题栏 返回按钮
                this.finish();
            }
            public void linshi(View v) { // 标题栏
                Utils.startAnim(mImgUp, mImgDn);
            }
        }
```

（5）点击“返回按钮”实现退出功能，点击“…”按钮，实现动画效果 CORE1108 所示。

代码 CORE1108:退出与动画效果

```
// 退出动画效果
public void shake_activity_back(View v) { // 标题栏 返回按钮
        this.finish();
    }
    public void linshi(View v) { // 标题栏
        Utils.startAnim(mImgUp, mImgDn);
    }
```

Android 的特色之一就是支持传感器,通过传感器可以获取手机运行的外界信息,包括手机运动的加速度、摆放方向等。学习本项目需要重点掌握 Android 传感器支持的 API,包括如何通过 SensorManager 注册传感器监听器等。除此之外,读者还需掌握其他传感器的使用方法。

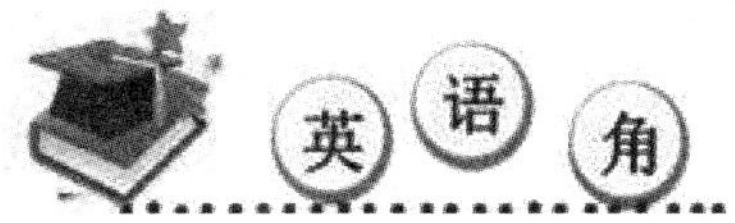

manager	管理器
count	计数
client	客户
extends	扩展
graphic	图像
host	主机
byte	字节
buffer	缓冲器
resolve	解析

一、选择题

1. 下列 LocationManager 获取位置信息途径的说法不正确的是(　　)。

A.GPS 定位更精确,缺点是只能在户外使用

B.NETWORK 通过基站和 Wi- Fi 信号来获取位置信息,速度较慢,耗电较少

C. 获取用户位置信息,我们可以使用其中一个,也可以同时使用两个

D.GPS 定位耗电严重,并且返回用户位置信息的速度远不能满足用户需求

2. 下列传感器可以用于制作微博里的“摇一摇”功能（即振动手机）来寻找周围同上微博人的是（　　）。

A.Sensor.TYPE_ORIENTATION　　B.Sensor.TYPE_PROXIMITY

C.Sensor.TYPE_ACCELEROMETER　　D.Sensor.TYPE_LIGHT

3. 使用地图不需要的权限是（　　）。

A.android.permission.ACCESS_WIFI_STATE

B.android.permission.WRITE_SECURE_SETTINGS

C.android.permission.INTERNET

D.android.permission. CHANGE_WIFI_STATE\

4. GoogleMap API 中用于控制地图的移动、缩放等功能的类是（　　）。

A.MapActivity　　B.MapView　　C.MapController　　D.GeoPoint

5. 在 Android 定位跟踪中，既提供访问定位服务的功能，又提供获取最佳定位提供者的功能的类是（　　）。

A.LocationManager　　B.LocationProvider

C.LocationListener　　D.Criteria

二、填空题

1.________ 接口是定义了常见的 Provider 状态变化和位置变化的方法。

2. 可以用来辅助 WebView 设置其一些属性和状态的类是 ________。

3. 要注册各种传感器需要先获取 ________ 对象。

4.Android 所有的传感器都归传感器管理器 ________ 管理。

5. ________ 是一个当 GPS 状态发生改变时，用来接收通知的接口。

三、判断题

1.Android 的定位方式有 GPS、通过网络的方式、基于基站的方式。（　　）

2.android.hardware.Camera.ShutterCallback 是 Camera 中处理快门关闭的接口。（　　）

3. 获取 LocationManager 的方法是：LocationManagerlm = (LocationManager) getSystemService(Context.LOCATION_SERVICE)。（　　）

4. 使用 HttpURLConnection 的 Get 方式请求数据时，connection.setDoInput(true) 必须设置。（　　）

5.AppWidget 窗口小部件时不可以使用 RelativeLayout。（　　）

四、简答题

1. GL 的坐标系和手机坐标系有什么区别？

2. 传感器包括哪些？

五、上机题

制作一个可以定位的地图 App（可以使用百度的类包），可以将手机处于移动状态显示出来,并且在地图上记录它的移动轨迹。

GetVersion()　　　　　　设备版本号

三个接口:

GpsStatus.Listener:这是一个当 GPS 状态发生改变时,用来接收通知的接口。

GpsStatus.NmeaListener:这是一个用来从 GPS 里接收 Nmea-0183(为海用电子设备制定的标准格式)信息的接口。

LocationListener:位置监听器,用于接收当位置信息发生改变时,从 LocationManager 接收通知的接口。

项目十二　网络编程

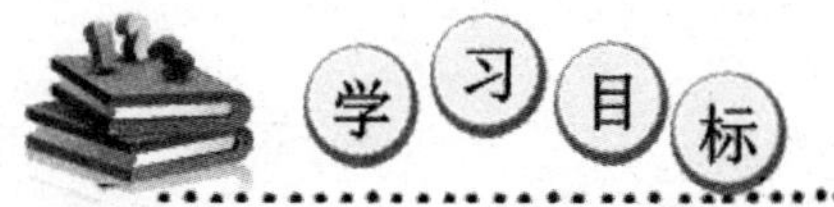

通过“火情监测系统”项目的实现，学习网络编程的相关知识，了解 Android 移动应用程序线程设计、Socket 通信、HTTP 通信的使用方法。在项目实现过程中：

- 掌握使用线程编程；
- 掌握 Socket 通信；
- 掌握 HTTP 通信；
- 掌握 Handler-Message 消息传递机制。

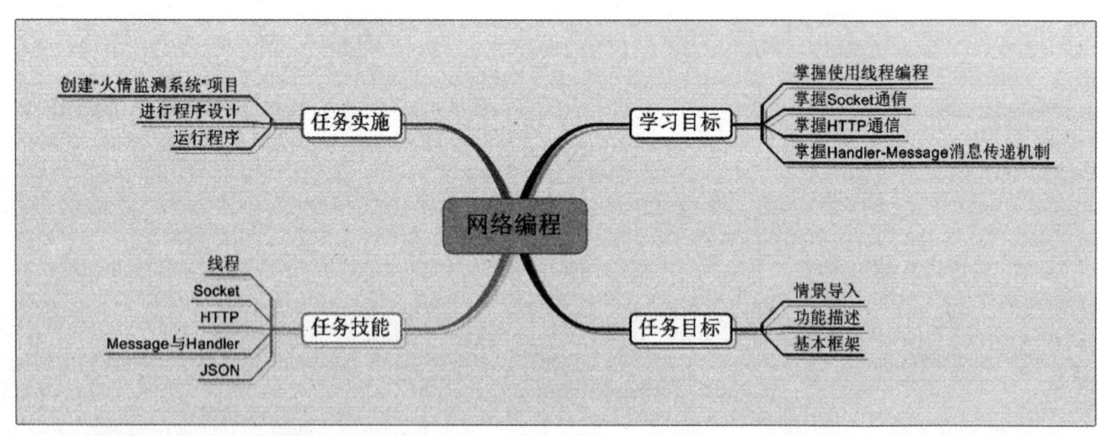

【情境导入】

日常生活中通过手机进行资料查询、交流、数据监控已成为大多数人的选择。本项目以火灾监测系统为背景，通过 Socket，HTTP 请求网络编程，实现火情监控系统中的核心功能，火灾信息监控以及回馈报警信息等。

【功能描述】

本项目将设计一款获取火情信息功能并且能够反馈信息的程序。

- 使用线性布局技术来设计登录系统界面；
- 点击“开始获取”按钮，建立 Socket 网络通信，并且开始获取火情数据显示到界面上；
- 点击“回馈信息”按钮，将火情信息情况回馈到服务端。

【基本框架】

基本框架如图 12.1 所示，将框架图转换成的效果如图 12.2 所示。

图 12.1　火情监测系统框架图

图 12.2　火情监测系统主界面效果图

技能点 1　线程

1　线程简介

线程是 CPU 调度和分派的基本单位，线程必须依赖进程而存活，并和其他线程共享依赖的进程的资源。线程是独立运行（相对于其他线程而言），线程也需要有自己的资源，包括栈、寄存器、状态、程序计时器。线程有新建、就绪、运行、阻塞、死亡五种状态。

2 线程实现方法

在 Android 中有实现线程 thread 的方法有两种，第一种是扩展 java.lang.Thread 类，第二种是实现 Runnable 接口。Thread 类代表线程类，它的两个主要方法是：run() 和 start()。

两种线程实现方法分别如下。

（1）java.lang.Thread 类，把 run() 方法写到线程中。

```
Thread thread = new Thread(new Runnable() {// 创建线程，新建 run 方法
@Override
public void run() {
// TODO Auto-generated method stub
while(true){
System.out.println("1111");
}
}}).start();// 开启线程
```

（2）实现 Runnable 接口，把 run() 方法单独提出。

```
Runnable r=new Runnable(){
   public void run() {// 创建 run 方法
      while(true){
         System.out.println("1111");
      } }
}
Thread t = new Thread(r);// 将 run 方法
t.start();// 开启线程
```

3 进程与线程的区别

在 Android 平台上一个程序（应用 App）是一个进程，一个进程至少有一个线程。线程的划分尺度小于进程，使得多线程程序的并发性高。进程在执行过程中拥有独立的内存单元，而多个线程共享内存，从而极大地提高了程序的运行效率。线程与进程在执行过程中的区别为：每个独立的线程都有一个程序运行的入口、顺序执行序列和程序的出口。线程不能够独立执行，必须依存在应用程序中，由应用程序提供多个线程执行控制。线程有自己的堆栈和局部变量，但线程之间没有单独的地址空间，一个线程死掉就等于整个进程死掉，所以多进程程序比多线程程序健壮。进程有独立的地址空间，一个进程崩溃后，在保护模式下不会对其他进程产生影响，而线程只是一个进程中的不同的执行路径。对于一些要求同时进行并且又要共享某些变量的并发操作，只能用线程，不能用进程。

拓展：想了解或学习更多进程知识，可扫描下方二维码，获取更多信息。

技能点 2　Socket

1　Socket 简介

工作于 TCP/IP 协议中应用层和传输层之间的一种抽象层为 Socket，在 Android 系统中，可以分为流套接字（streamsocket）和数据报套接字（datagramsocket）。而 Socket 中的流套接字将 TCP 协议作为其端对端协议，提供了一个可信赖的字节流服务，数据报套接字使用 UDP 协议，提供数据打包发送服务。Socket 工作机制中包括服务端和客户端两部分。在服务端有多个端口，每个端口由端口号标识。当客户端与服务端建立连接时，首先服务端打开端口监听来自客户端的请求，然后客户端通过 IP 地址和端口号向服务端发送连接请求，最后服务端接收请求，若连接成功，可以开始通信。

注意：应用程序使用 TCP/IP 协议进行通信，须通过 Socket 与操作系统交互并请求服务。

2　Socket 实现方法

当客户端、服务器端产生了对应的 Socket 之后，程序无须再区分服务器、客户端，而是通过各自的 Socket 进行通信。Socket 提供了两个方法获取输入流和输出流，如表 12.1 所示。

表 12.1　Socket 获取流方法

方法	说明
getInputStream()	返回该 Socket 对象对应的输入流，让程序通过该输入流从 Socket 中取出数据
getOutputStream()	返回该 Socket 对象对应的输出流，让程序通过该输出流向 Socket 中输出数据

Socket 通信步骤：

（1）遍历 Socket 实现通信连接。

```
Socket socket = new Socket(url,port);
```

（2）获取输入流，用户可根据需求将流转换成所需格式。

```
IntputStream stream  = socket.getInputStream();
```

（3）Android 实现 Socket 简单通信需要添加的权限。

```
<!-- 允许应用程序改变网络状态 -->
<uses-permission android:name="android.permission.CHANGE_NETWORK_STATE"/>
<!-- 允许应用程序改变 WIFI 连接状态 -->
<uses-permission android:name="android.permission.CHANGE_WIFI_STATE"/>
<!-- 允许应用程序访问有关的网络信息 -->
<uses-permission android:name="android.permission.ACCESS_NETWORK_STATE"/>
<!-- 允许应用程序访问 WIFI 网卡的网络信息 -->
<uses-permission android:name="android.permission.ACCESS_WIFI_STATE"/>
<!-- 允许应用程序完全使用网络 -->
<uses-permission android:name="android.permission.INTERNET"/>
```

技能点 3　HTTP

1　HTTP 简介

超文本传输协议（Hypertext Text Transfer Protocol，简称 HTTP）是应用层协议，自 1990 年起，HTTP 就已经被应用于 WWW 全球信息服务系统。HTTP 是一种请求 / 响应式的协议。一个客户机与服务器建立连接后，会发送一个请求给服务器，这个服务器接到请求后，会给予客户机相应信息。

HTTP 的第一版是一种简单的用于网络间原始数据传输的协议，HTTP/1.0 由 RFC1945 定义，进一步的改进 HTTP/0.9，它允许消息是类 MIME 信息格式。

2　HTTP 请求

HTTP 包含了两种请求方式：GET 和 POST。GET 请求一般用于获取或查询资源信息，POST 请求一般用于更新资源信息。HttpGet、HttpPost 分别实现了 HttpRequest、HttpUriRequest 接口，构造方法如表 12.2 和表 12.3 所示。

表 12.2　HttpGet 构造方法

名称	说明
public HttpGet()	无参数构造方法用以实例化对象
public HttpGet(URI uri)	通过 URI 对象构造 HttpGet 对象
public HttpGet(String uri)	通过指定的 uri 字符串地址构造实例化 HttpGet 对象

表 12.3　HttpPost 构造方法

名称	说明
public HttpPost()	无参数构造方法用以实例化对象
public HttpPost (URI uri)	通过 URI 对象构造 HttpPost 对象
public HttpPost (String uri)	通过指定的 uri 字符串地址构造实例化 HttpPost 对象

Android 提供了 HttpURLConnection 和 HttpClient 接口来开发 HTTP 程序。HttpURLConnection 是 Java 的标准类，继承自 HttpConnection。它是抽象类，不能实例化对象，主要通过 URL 的 openConnection() 方法获得。HttpConnection 接口的常用抽象方法如表 12.4 所示。

表 12.4　HttpConnection 接口的常用抽象方法

名称	说明
conn.setDoInput(true)	设置输入流
conn.setDoOutput(true)	设置输出流
conn.setConnectTimeout(10000)	设置超时时间
conn.setRequestMethod("GET")	设置请求方式（GET 或 POST）
conn.setUseCaches(false)	POST 请求不能使用缓存

3　HttpClient 开发

Apache 提供了一个 HttpClient 项目，能更好地处理向 Web 站点请求，包括处理 Session、Cookie 等细节问题。Apache 是一个简单的 HTTP 客户端，可以发送 HTTP 请求，接收 HTTP 响应，执行过程如图 12.3 所示，但不会缓存服务器的响应，HttpClient 接口的常用抽象方法如表 12.5 所示。

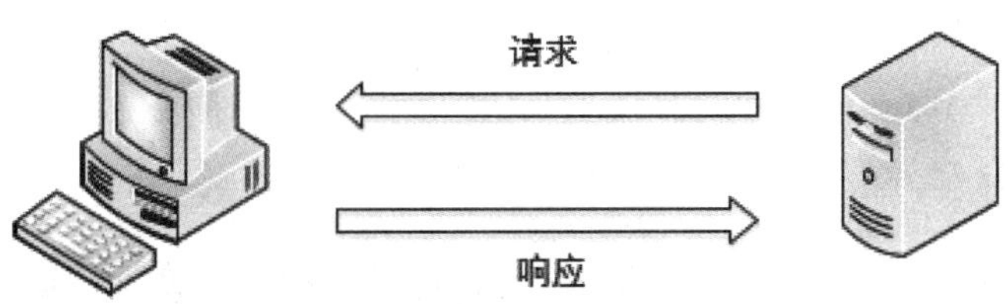

图 12.3　请求与响应执行过程

表 12.5　HttpClient 接口的常用抽象方法

名称	说明
public abstract HttpResponse execute(HttpUriRequest request)	通过 HttpUriRequest 对象执行返回一个 HttpResponse 对象
public abstract HttpResponse execute(HttpUriRequest request,HttpContext context)	通过 HttpUriRequest 对象和 HttpContext 对象执行返回一个 HttpResponse 对象

使用 HttpClient 发送请求、接收请求步骤如下：

（1）创建一个 HttpClient 对象。

```
HttpClient client = new HttpClient();
```

（2）若需要发送 GET 请求，则创建 HttpGet 对象；若需要发送 POST 请求，则创建 HttpPost 对象。

GET 请求方法：

```
HttpGet get = new HttpGet("http://www.baidu.com");
```

POST 请求方法：

```
HttpPost post = new HttpPost("http://localhost/....");
```

（3）调用 HttpClient 对象的 execute(HttpUriRequest request) 方法来发送请求，执行过后该方法返回一个 HttpResponse。

```
httpResponse = httpClient.execute(get);
```

（4）调用 HttpResponse 的 getEntity() 方法可获取 HttpEntity 对象，这个对象包装了服务器的响应内容。

```
HttpEntity entity = httpResponse.getEntity();
```

技能点 4 Message 与 Handler

1 Message 简介

当 Android 平台启动一个应用程序时会开启一个主线程（界面 UI 线程），界面 UI 线程管理显示的所有控件，并监听用户点击事件响应用户分发事件。在界面 UI 线程中一般不执行耗时的操作，如联网下载数据等，会出现 ANR 错误。AndroidUI 线程是不安全的，所以只能在主线程中更新 UI。

正如其他 GUI 应用程序一样，Android 应用程序也是消息（事件）驱动的。这种消息的传递必须依赖于应用框架提供的消息机制。Android 本身提供了两种消息机制：组件间消息传递（Intent）和线程间消息传递（Message）。在此主要讨论 Android 线程间消息传递机制及其应用。

android.os.Message 是定义一个 Message 包含的必要的描述和属性数据，并且此对象可以被发送给 android.os.Handler 处理。属性字段：arg1、arg2、what、obj、replyTo 等；其中 arg1 和

arg2 是用来存放整型数据的；what 是用来保存消息标识；obj 是 Object 类型的任意对象；replyTo 是消息管理器，会关联到一个 Handler，Handler 处理其中的消息。通常 Message 对象不是直接 new 出来的，调用 Handler 中的 obtainMessage 方法获得 Message 对象如表 12.6 所示。

表 12.6　获得 Message 对象方法

方法	含义
Message()	构造一个新的 Message 对象
Message obtain ()	从全局池中返回一个新的 Message 实例

2　Handler 简介

使用 Handler 处理消息时，Handler 发挥的两个作用分别是：在新线程中发送消息，在主线程（界面线程）中获取并处理消息。在实际的程序开发中，Handler 类为开发人员提供了便捷的开发策略，在新线程（可以是多个新线程）中编写消息发送的功能代码，在主线程中统一接收、识别并处理。解决何时发送、何时处理的问题。常用的一些方法说明如表 12.7 所示。

表 12.7　常用方法说明

方法	含义
void　handleMessage(Message msg)	主线程处理消息的方法
boolean　hasMessages(int what, Object object)	检查消息队列是否包含有 what 属性并且 object 属性为指定对象的消息
final boolean　hasMessages(int what)	检查消息队列是否包含有 what 属性的消息
final Message　obtainMessage()	获取消息
final boolean　post(Runnable r)	将一个线程添加到消息队列
final boolean　postDelayed(Runnable r, long delayMillis)	将一个线程 delayMillis 毫秒后添加到消息队列
final boolean　sendMessage(Message msg)	立即发送消息
final boolean　sendMessageDelayed(Message msg, long delayMillis)	延迟 delayMillis 毫秒后发送消息

在开发移动应用程序时，开发人员只需要重写 Handler 类中的处理消息的 handleMessage(Message msg) 方法，每当新线程使用 sendMessage(Message msg) 发送消息时，Handler 类会自动回调 handleMessage(Message msg) 的逻辑代码。Handler 用法流程如图 12.4 所示。

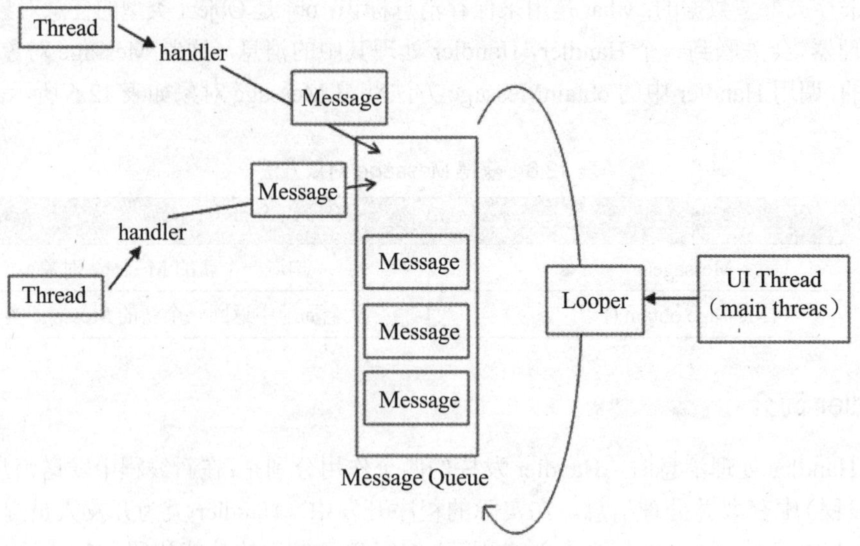

图 12.4 Handler 流程图

3 Handler-Message 消息传递机制

使用 Handler-Message 消息传递机制更新主线程 UI 步骤如下：

（1）在主线程 Activity 中创建 Handler 对象。

```
// 创建 Handler 方法
Handler mHandler = new Handler(){
        public void handleMessage(android.os.Message msg) {
}
    };
```

（2）在新线程中使用主线程创建的 Handler 对象，调用它的发送消息方法向主线程发送消息。

```
//Message 发送消息机制
Message mMsg = new Message();// 创建 Message 对象
mMsg.obj = 0;// 设置 mMsg 的传输值
mMsg.what =str; // 设置 mMsg 的类型
mHandler.sendMessage(mMsg); // 通过构造函数传入的 Handler 发送 Message
```

（3）利用 Handler 对象的 handleMessage(Message msg) 方法接收消息，然后根据 obj 的不同取值执行不同的业务逻辑。

```
Handler mHandler = new Handler(){
    public void handleMessage(android.os.Message msg) {
```

```
switch (msg.what) {
    case 0:
String str = (String)msg.obj;
    break;
        }
            };
    };
```

技能点 5　JSON

1　JSON 简介

JSON 是一种轻量级的数据交换格式，完全独立于文本格式，易于读者阅读和编写，同时也易于解析和生成。服务器请求成功后得到的数据大多是 JSON 类型的数据，而不是客户所需要的明确信息。该数据有特定的结构格式和语义，JSON 格式如下所示：


```
[{"title":" 不 要 这 样 呢 ","thumburl":"http://ww3.sinaimg.cn/large/bd759d6djw1f5z-1s16t0cj20b70ci74x.jpg","sourceurl":"http://down.laifudao.com/images/tupi-an/2016717103538.jpg","height":"45  0","width":"403","class":"1","url":"http://www.laifud-ao.com/tupian/59837.htm"},{"title":" 女汉子的 ","thumburl":"http://ww2.sinaimg.cn/large/e4e2bea6jw1f5r72w5bayj20be0a7jru.jpg","sourceurl":"http://down.laifudao.com/images/tu-pian/201676173157.jpg","height":"367","width":"410","class":"1","url":"http://www.laifud-ao.com/tupian/59578.htm"},{"title":"haha    去 ","thumburl":"http://ww2.sinaimg.cn/large/bd698b0fjw1f58fyhmi75j20ge0j675p.jpg","sourceurl":"http://down.laifudao.com/images/tu-pian/2016619142910.jpg","height":"690","width":"590","class":"1","url":"http://www.laifud-ao.com/tupian/58944.htm"}]
```


2　JSON 解析方式

JSON 解析一般有三种方式，原生解析方式，JSON 解析方式和 FastJson 解析方式。本次讲解原生解析方式，步骤如下。

第一步：将得到的数组 JSON 进行拆分，拆分为字符串形式。

```
    try {
// 创建 Json 数组对象
                JSONArray  array = new JSONArray(json);
```

```
        } catch (JSONException e) {
            // TODO Auto-generated catch block
            e.printStackTrace();
        }
```

第二步：将字符串形式的 JSON 进行进一步拆分，将其中各个对象取出。

```
    try {
    // 解析 Json 数组，将其转换为 Json 字符，最后进行解析
                for (int i = 0; i < array.length(); i++) {
                JSONObject object = array.getJSONObject(i);
                String title = object.getString("title");// 获取 key
                String thumburl = object.getString("thumburl");
                String sourceurl = object.getString("sourceurl");
                String height = object.getString("height");
                String width = object.getString("width");
                String ass = object.getString("class");
                String url = object.getString("url");
            }
        } catch (JSONException e) {
            // TODO Auto-generated catch block
            e.printStackTrace();
        }
```

第一步：在 Eclipse 中创建一个 Android 工程，命名为"火情监测系统"，并设计界面。如图 12.2 所示。

第二步：在 src 文件夹中下建立 MainActivity 文件，并实现点击"开始获取"按钮，建立 Socket 通信，获取火情信息显示到界面上。具体如代码 CORE1201 所示。

代码 CORE1201：建立通信，发送信息

```
/**
*1 此处添加 Socket 连接获取流，并将信息抛出功能代码
*/
button.setOnClickListener(new OnClickListener() {
        @Override
```

```
            public void onClick(View arg0) {
                // TODO Auto-generated method stub
                getdata();
            }
        });
    protected void getdata() {
        // TODO Auto-generated method stub
        new Thread(new Runnable() {
            @SuppressWarnings("static-access")
            @Override
            public void run() {
                // TODO Auto-generated method stub
                try {
                    s = new Socket("192.168.1.102", 9988);
                    System.out.println("Socket 成功 ");
                    while (true) {
                    byte[] bs = new byte[1024];
                        InputStream inputStream = s.getInputStream();
                        int len = inputStream.read(bs);
                        System.out.println(bs[0] + "---------------");
                        if (len > 0) {
                            if (bs[0] == 1) {
                                text = " 火焰:有 ";
                                Message msg = new Message();
                                msg.obj = text;
                                msg.what = 0;
                                handler.sendMessage(msg);
                            }
                        }
                        Thread.currentThread().sleep(200);
                    }
                } catch (Exception e) {
                    // TODO Auto-generated catch block
                    e.printStackTrace();
                }
            }
        }).start();
    }
```

第三步：接收抛出信息，并显示到界面上，具体如代码 CORE1202 所示。

代码 CORE1202：建立通信，接收信息

```
/**
*2 此处添加接收有 Message 抛出的信息
*/
Handler handler = new Handler() {
        public void handleMessage(Message msg) {
            String str = (String) msg.obj;
            textView.setTextColor(Color.RED);
            textView.setText(str);
        };
    };
```

第四步：实现点击“回馈信息”按钮，将火灾情况回馈到服务端并显示到界面上。具体如代码 CORE1203 所示。

代码 CORE1203：回馈信息

```
/**
*3 此处添加信息回馈代码
*/
button2.setOnClickListener(new OnClickListener() {
            @Override
            public void onClick(View arg0) {
                // TODO Auto-generated method stub
                try {
                    String str = " 已接收报警信息 ";
                    byte[] bs2 = str.getBytes();
                    OutputStream outputStream = s.getOutputStream();
                    outputStream.write(bs2, 0, bs2.length);
                    outputStream.close();
                    textView2.setText(" 回馈信息："+str);
                } catch (IOException e) {
                    e.printStackTrace();
                }
            }
        });
```

第五步：运行程序，运行结果如图 12.5 所示 。

图 12.5　火情监测系统主界面

【拓展目的】

熟悉并掌握使用线程实现网络编程获取信息。

【拓展内容】

本任务设计一款获取服务器图片的软件，效果如图 12.6 所示。

图 12.6　拓展主界面

【拓展步骤】

（1）设计思路：点击“获取图片”按钮，获取服务器图片，将获取到的图片显示到界面。

（2）获取服务器图片。具体如代码 CORE1204 所示。

代码 CORE1204: 获取服务器图片

```
/**
*1 此处填写获取服务器图片代码
*/
public class MainActivity extends Activity {
    Button button;
    ImageView imageView;
        String url = "http://192.168.1.102:1234/Handler1.ashx";
        @Override
        protected void onCreate(Bundle savedInstanceState) {
                super.onCreate(savedInstanceState);
                setContentView(R.layout.activity_main);
                button = (Button) findViewById(R.id.button1);
                imageView = (ImageView) findViewById(R.id.imageView1);
                button.setOnClickListener(new OnClickListener() {
                        @Override
                        public void onClick(View arg0) {
                                // TODO Auto-generated method stub
                                new Thread(new Runnable() {
                                        public void run() {
                                                // TODO Auto-generated method stub
                                                try {
                                                        HttpClient client=new DefaultHttpClient();
                                                        HttpPost httpPost=new HttpPost(url);
                                                        httpPost.setEntity(new StringEntity("", "utf-8"));
                                                            HttpResponse response=client.execute(httpPost);
                                        System.out.println(response.getStatusLine().getStatusCode
()+"-----------");
                                                            if(respons.getStatusLine().getStatusCode()==
200){
                                                        final  InputStream  stream=response.getEntity().get-
Content();
                                                       Bitmap bitmap = BitmapFactory.decodeStream(stream);
                                                                        Message msg = new Message();
                                                                        msg.obj=bitmap;
                                                                        msg.what=0;
                                                                        handler.sendMessage(msg);
                                                            }
```

```
                                } catch (Exception e) {
                                    // TODO: handle exception
                                }
                        }
                    }).start();
                }
            });
        }
    }
```

（3）更新界面图片。具体如代码 CORE1205 所示。

代码 CORE1205：更新界面图片

```
/**
*2 此处添加更新界面图片功能代码
*/
Handler handler = new Handler(){
        public void handleMessage(Message msg) {
            switch (msg.what) {
            case 0:
                Bitmap bitmap = (Bitmap) msg.obj;
                System.out.println(bitmap+"----Bitmap-------");
                imageView.setImageBitmap(bitmap);
                break;
            default:
                break;
            }
        };
    };
```

本项目主要介绍了 Android 网络编程的相关知识。Android 支持 JDK 网络编程中的 ServiceSocket、Socket 等 API，也支持 JDK 内置的 URL、URLConnection、HttpURLConnection 等工具类，通过本项目的学习，读者要掌握网络编程的基本知识，并且熟练运用到以后的代码编写中。

response 响应
request 请求
service 服务
destroy 销毁
startup 启动
read 读取
NULL PointerException 空指针异常
finally 最后
listener 收听者
exception 异常

一、选择题

1. 使用 HttpClient 的 Get 方式请求数据时，可以用（　　）类来构建 Http 请求。

A.Get　　B.URLConnection　　C.HttpGet　　D.HttpPost

2. 下列关于 Socket 通讯正确的说法是（　　）。

A. 服务器端需要 ServerSocket 需要绑定端口号

B. 服务器端需要 ServerSocket 需要绑定端口和 IP 地址

C. 客户端需要 Socket，需要绑定端口号

D. 客户端需要 ServerSocket，需要绑定端口号

3. 在 Web Service 中，通过（　　）来执行服务调用。

A.HTTPS　　B.SOAP　　C.XML Schema　　D.UDDI

4. Thread 类中，能运行线程体的方法是（　　）。

A.Start()　　B.Resume()　　C.init()　　Drun()

5. 下列关于线程说法不正确的是（　　）。

A. 在 Android 中，我们可以在主线程中创建一个新的线程

B. 在创建的新线程中，它可以操作 UI 组件

C. 新线程可以和 Handler 共同使用

D. 创建的 Handler 对象，它隶属于创建它的线程

二、填空题

1. 一个 Handler 允许发送和处理 Message 或者 Runnable 对象，并且会关联到主线程的_______中。

2. 应用 Socket 创建 ___ 对象，用于接收 _______ 中的数据。

3. 在 Android 开发中，Android SDK 附带了 Apache 的 HttpClient，它是一个完善的客户端。它提供了对 HTTP 协议的全面支持，可以使用 HttpClient 的对象来执行 _______ 和 _______ 调用。

4. Web Service 编程中，创建 SoapObject 对象时需要传入所要调用 Web Service 的 _______、Web Service 方法名；如果有参数要传给 Web Service 服务器，调用 _______ 方法来设置参数，该方法的 name 参数指定参数名。

5. 调用 HttpResponse 的 _______、_______ 等方法可获取服务器的响应。

三、判断题

1.WebService 是一种基于 HTTP 协议的远程调用标准。（　）

2. socket 通信中，创建 HttpTransportSE 传输对象，传入 WebService 服务器地址就能实现通信。（　）

3. Handler 主要有两个作用：在工作线程中发送消息和在 UI 线程中获取、处理消息。（　）

4. 设置与 .NET 提供的 Web Service 保持有良好的兼容性：SoapSerializationEnelope 类的对象实例名 .dotNet = true。（　）

5. POST 和 GET 请求没有区别。（　）

四、简答题

1. 简述 Android 调用 WebService 的步骤。

2. 简述 httpclient 通信中以 GET 方法和 POST 方法的缺点。

五、上机题

编写程序获取 Tomcat 服务器下的文件数据。